Other Geographies

Antipode Book Series

Series Editors: Vinay Gidwani, University of Minnesota, USA and Sharad Chari, University of California, Berkeley, USA

Like its parent journal, the Antipode Book Series reflects distinctive new developments in radical geography. It publishes books in a variety of formats – from reference books to works of broad explication to titles that develop and extend the scholarly research base – but the commitment is always the same: to contribute to the praxis of a new and more just society.

Published

Other Geographies: The Influences Of Michael Watts
Edited by Sharad Chari, Susanne Freidberg, Vinay Gidwani, Jesse Ribot and Wendy Wolford

Money and Finance After the Crisis: Critical Thinking for Uncertain Times
Edited by Brett Christophers, Andrew Leyshon and Geoff Mann

Frontier Road: Power, History, and the Everyday State in the Colombian Amazon
Simón Uribe

Enterprising Nature: Economics, Markets and Finance in Global Biodiversity Politics
Jessica Dempsey

Global Displacements: The Making of Uneven Development in the Caribbean
Marion Werner

Banking Across Boundaries: Placing Finance in Capitalism
Brett Christophers

The Down-deep Delight of Democracy
Mark Purcell

Gramsci: Space, Nature, Politics
Edited by Michael Ekers, Gillian Hart, Stefan Kipfer and Alex Loftus

Places of Possibility: Property, Nature and Community Land Ownership
A. Fiona D. Mackenzie

The New Carbon Economy: Constitution, Governance and Contestation
Edited by Peter Newell, Max Boykoff and Emily Boyd

Capitalism and Conservation
Edited by Dan Brockington and Rosaleen Duffy

Spaces of Environmental Justice
Edited by Ryan Holifield, Michael Porter and Gordon Walker

The Point is to Change it: Geographies of Hope and Survival in an Age of Crisis
Edited by Noel Castree, Paul Chatterton, Nik Heynen, Wendy Larner and Melissa W. Wright

Privatization: Property and the Remaking of Nature-Society
Edited by Becky Mansfield

Practising Public Scholarship: Experiences and Possibilities Beyond the Academy
Edited by Katharyne Mitchell

Grounding Globalization: Labour in the Age of Insecurity
Edward Webster, Rob Lambert and Andries Bezuidenhout

Decolonizing Development: Colonial Power and the Maya
Joel Wainwright

Cities of Whiteness
Wendy S. Shaw

Neoliberalization: States, Networks, Peoples
Edited by Kim England and Kevin Ward

The Dirty Work of Neoliberalism: Cleaners in the Global Economy
Edited by Luis L. M. Aguiar and Andrew Herod

David Harvey: A Critical Reader
Edited by Noel Castree and Derek Gregory

Working the Spaces of Neoliberalism: Activism, Professionalisation and Incorporation
Edited by Nina Laurie and Liz Bondi

Threads of Labour: Garment Industry Supply Chains from the Workers' Perspective
Edited by Angela Hale and Jane Wills

Life's Work: Geographies of Social Reproduction
Edited by Katharyne Mitchell, Sallie A. Marston and Cindi Katz

Redundant Masculinities? Employment Change and White Working Class Youth
Linda McDowell

Spaces of Neoliberalism
Edited by Neil Brenner and Nik Theodore

Space, Place and the New Labour Internationalism
Edited by Peter Waterman and Jane Wills

Other Geographies

The Influences Of Michael Watts

Edited by

Sharad Chari, Susanne Freidberg, Vinay Gidwani,
Jesse Ribot and Wendy Wolford

WILEY Blackwell

Registered Office(s)
John Wiley & Sons, Inc., 111 River Street, Hoboken, NJ 07030, USA
John Wiley & Sons Ltd, The Atrium, Southern Gate, Chichester, West Sussex, PO19 8SQ, UK

Editorial Office
9600 Garsington Road, Oxford, OX4 2DQ, UK

For details of our global editorial offices, customer services, and more information about Wiley products visit us at www.wiley.com.

Wiley also publishes its books in a variety of electronic formats and by print-on-demand. Some content that appears in standard print versions of this book may not be available in other formats.

Library of Congress Cataloging-in-Publication data applied for

9781119184768 [hardback]
9781119184775 [paperback]

Cover image: Bucket Heads © Sokari Douglas Camp. All rights reserved, DACS 2017.
Cover design: Wiley

Set in 10.5/12.5pt Sabon by SPi Global, Pondicherry, India
Printed and bound in Malaysia by Vivar Printing Sdn Bhd

10 9 8 7 6 5 4 3 2 1

Contents

Series Editors' Preface

The *Antipode Book Series* explores radical geography 'antipodally', in opposition, from various margins, limits, or borderlands.

Antipode books provide insight 'from elsewhere', across boundaries rarely transgressed, with internationalist ambition and located insight; they diagnose grounded critique emerging from particular contradictory social relations in order to sharpen the stakes and broaden public awareness. An *Antipode* book might revise scholarly debates by pushing at disciplinary boundaries, or by showing what happens to a problem as it moves or changes. It might investigate entanglements of power and struggle in particular sites, but with lessons that travel with surprising echoes elsewhere.

Antipode books will be theoretically bold and empirically rich, written in lively, accessible prose that does not sacrifice clarity at the altar of sophistication. We seek books from within and beyond the discipline of geography that deploy geographical critique in order to understand and transform our fractured world.

Vinay Gidwani
University of Minnesota, USA

Sharad Chari
University of California, Berkeley, USA

Antipode Book Series Editors

Notes on Contributors

Teo Ballvé is an Assistant Professor in Peace & Conflict Studies and Geography at Colgate University in upstate New York. Before returning to academia, he worked for many years as a journalist, covering Latin American affairs and US policy toward the region. He is the co-editor (with Vijay Prashad) of *Dispatches from Latin America: On the Frontlines Against Neoliberalism* (South End Press; LeftWord, 2006).

Thomas J. Bassett's research centres on the political ecology of agrarian change in West Africa. He earned his PhD in Geography at the University of California, Berkeley, in 1984. He was Professor Michael Watts's first PhD student. Dr Bassett's long-term research in West Africa grapples with the question of why peasant farmers and herders, despite their access to land and labour, remain vulnerable to food insecurity. His research in Côte d'Ivoire traces the transformation of farming and pastoral systems, their interactions with markets and the state, and the multi-scale political ecological dynamics that produce vulnerability as well as opportunities for reducing it. His recent publications focus on world market prices and cotton grower incomes in West Africa (*World Development*), the adaptation concept in the climate change literature (*Geoforum*) and political ecological perspectives on socio-ecological relations (*Natures, Sciences et Sociétés*). He also writes on the history of maps and mapmaking in Africa with contributions to three volumes of the six-volume *The History of Cartography*.

Joe Bryan's research draws from 20 years of experience working with indigenous movements in the Americas, including work in Ecuador, Nicaragua, the United States and Mexico. Much of that work focuses on efforts by indigenous peoples to formulate claims to territory, in particular through the production of maps. He has written extensively on his work in both English and Spanish, and is the co-author, with Denis Wood, of *Weaponizing Maps: Indigenous Peoples and Counterinsurgency in the Americas*. He is currently Associate Professor of Geography at the University of Colorado, Boulder.

Judith Carney is Professor of Geography at UCLA. Her research centres on African ecology and development, food security, gender and agrarian change, and African contributions to New World environmental history. She is the author of more than 90 scholarly articles and two books: *Black Rice: The African Origins of Rice Cultivation in the Americas* (Harvard University Press, 2001) and *In the Shadow of Slavery: Africa's Botanical Legacy in the Atlantic World* (University of California Press, 2009). *Black Rice* received the Melville Herskovits Book Award and *In the Shadow of Slavery*, the Frederick Douglass Book Prize. She has received professional honours from the Association of American Geographers, including a Distinguished Scholarship Honor, the Robert Netting Award for original research that bridges geography and anthropology and the Carl O. Sauer Distinguished Scholarship Award for significant contributions to Latin American geography. Her research has been supported by the National Geographic Society, the Guggenheim Foundation, the American Council of Learned Societies, the Rockefeller Foundation and the Wenner-Gren Foundation. Currently she is researching human use of West African mangrove ecosystems in the context of climate change and continuing her collaboration with plant scientists on historical and geographical themes concerning the genome sequencing of African rice.

Sharad Chari is at the Centre for Indian Studies in Africa and the Department of Social Anthropology at the University of the Witwatersrand, Johannesburg, South Africa, and, from 2017, at the Department of Geography at Berkeley. He has taught at the LSE and at Michigan, where he was in Anthropology, History, and the Michigan Society of Fellows. Sharad has worked on gender, caste and work politics in agrarian and industrial South India, in *Fraternal Capital: Peasant-workers, self-made men, and globalization in provincial India* (Stanford, 2004); development theories and trajectories, in the edited *Development Reader* (Routledge, 2008, with Stuart Corbrige); he is finishing a palimpsestic book on the past and present of racial capitalism and opposition in South Africa, called *Apartheid Remains*; and he is beginning research on archaic and emergent formations of racial/sexual capitalism in the Southern African Indian Ocean region. He works with agrarian studies, the Black radical tradition, documentary photography and other traditions of Earth-writing that have sought to stretch Marxist thought to realities considered (but not actually) peripheral to the planet.

Erin Collins is Assistant Professor of Global Urban Studies in the School of International Service at American University in Washington DC, USA. Dr Collins is a critical, urban geographer whose work focusses on the political economy and cultural politics of transformation in Southeast Asian cities. She received her PhD in Geography from the University of California, Berkeley in 2015. Her current research looks at how people claim and defend space in Cambodia's capital city of Phnom Penh in and through moments of political, social and economic remaking.

Rosalind Fredericks is Assistant Professor at New York University's Gallatin School of Individualized Study. After her PhD in Geography at the University of California, Berkeley, she was a postdoctoral research scholar with the Committee on Global Thought at Columbia University. Trained as an urban and cultural geographer, her research interests are centred on urban development, citizenship, political ecology, infrastructure and geographies of waste in Africa. Her forthcoming book, *Garbage Citizenship: Vibrant Infrastructures of Labor in Dakar, Senegal* (Duke University Press) chronicles the infrastructural politics surrounding municipal garbage labour in the wake of structural adjustment. A new research project funded by the National Science Foundation examines planning and activism surrounding the proposed closure of the city's dump, Mbeubeuss. She also has an ongoing research project on the role of hip hop in elections in Senegal. Fredericks has edited two books with Mamadou Diouf on citizenship in African cities, *Les Arts de la Citoyenneté au Sénégal: Espaces Contestés et Civilités Urbaines* (Editions Karthala, 2013) and *The Arts of Citizenship in African Cities: Infrastructures and Spaces of Belonging* (Palgrave MacMillan, 2014).

Susanne Freidberg received her PhD from Berkeley Geography in 1996 and is now Professor of Geography at Dartmouth College. Her research centres on the politics, practices and cultural meanings of food supply chains. While her dissertation examined the social and environmental history of commercial gardening in Burkina Faso, her more recent work focuses on the agricultural sustainability initiatives undertaken by the world's biggest food companies. She is the author of two books, *French Beans and Food Scares: Culture and Commerce in an Anxious Age* (Oxford, 2004) and *Fresh: A Perishable History* (Harvard, 2009), as well as articles that have appeared in journals such as *Economy and Society, Comparative Studies in Society and History, Science and Culture, Geoforum* and *Gastronomica*.

Benjamin Gardner is Associate Professor of Global Studies, Environmental Studies and Cultural Studies in the School of Interdisciplinary Arts and

Sciences at the University of Washington Bothell. He is also Chair of the African Studies Program in the Henry M. Jackson School of International Studies at the University of Washington. His research examines the relationship between tourism, conservation and development. He teaches courses on globalization, political ecology, and cultural studies theory and methods. His book, *Selling the Serengeti: The Cultural Politics of Safari Tourism* (Georgia, 2016) draws on cultural geography, environmental history and political economy to question the pervasive myths about who owns nature in Africa and how colonial discourses around conservation continue to shape contemporary environmental politics. He is a recipient of the University of Washington's Distinguished Teaching Award (2014). He has a BA in anthropology from Connecticut College, an MS in environmental studies from Yale University and a PhD in geography from the University of California Berkeley.

Vinay Gidwani is Professor of Geography and Global Studies at University of Minnesota. He studies the entanglements of labour and ecology in agrarian and urban settings, and capitalist transformations of these. He is particularly interested in the cultural politics and geographies of work. Vinay is the author of *Capital, Interrupted: Agrarian Development and the Politics of Work in India* (University of Minnesota Press, 2008). Recent publications include articles in *Transactions of the Institute of British Geographers*, *Comparative Studies of South Asia, Africa, and the Middle East* and *Economic and Political Weekly* (India). He is presently working on an ACLS-funded collaborative research project on the life-worlds of urban migrants in India who work in precarious informal economy jobs, and will be soon embarking on a new NSF-funded comparative study of Jakarta, Indonesia and Bangalore, India called 'Speculative Urbanism: Land, Livelihoods, and Finance Capital' with collaborators at Minnesota, UCLA, the National Institute of Advanced Study (Bangalore) and Tarumanagara University (Jakarta).

Julie Guthman is a geographer and Professor of Social Sciences at the University of California at Santa Cruz where she teaches courses primarily in global political economy and the politics of food and agriculture. She has published extensively on contemporary efforts to transform food production, distribution and consumption. Her publications include two multi-award-winning books: *Agrarian Dreams: the Paradox of Organic Farming in California* and *Weighing In: Obesity, Food Justice, and the Limits of Capitalism*. She is the recipient of the 2015 Excellence in Research Award from the Agriculture, Food and Human Values Society.

Lucy Jarosz is Chair and Professor of Geography at the University of Washington. She first met Michael Watts through his book, *Silent Violence*, a gift that her advisor, Harold Scheub, Professor of African Language and Literature at the University of Wisconsin-Madison, gave to her. That book was an important introduction to Geography. Michael's intellectual breadth and depth have remained an inspiration to her. She especially appreciates his interest in literature and the art of photography – in particular that of August Sander. Her interest in rural development, agriculture and food has remained constant since her graduate school days as one of his students at Berkeley. She has been fortunate to continue to study these topics in Madagascar, South Africa, the US and Canada. Her research draws from feminist political ecology and critical discourse analysis to examine how hunger and poverty are produced, addressed or magnified through agricultural development and change.

Moussa Koné earned his PhD in Geography from the University of Illinois Urbana-Champaign where he worked with Professor Thomas Bassett. He is Maître Assistant and teaches at the Institut de Géographie Tropicale (IGT), University Félix Houphouet-Boigny, Cocody-Abidjan, Côte d'Ivoire. Dr Koné's research interests centre on the political ecology of natural resource management and land rights systems. He utilizes geospatial technologies and qualitative and quantitative field techniques to investigate the social and biophysical dimensions of environmental change with emphasis on vulnerability and adaptation to climate change and variability. Dr. Koné currently participates in research projects that (1) assess the impact of the value chain approach on women farmers and household food security in the context of the New Green Revolution for Africa and (2) investigate how farmers and herders use fire as a tool for natural resource management in West African savannas, and how these practices modify landscapes over time and contribute to greenhouse gas emissions.

Jake Kosek is Associate Professor in Geography at Berkeley. He has been a Lang Postdoctoral Fellow at Stanford and a Ciriacy-Wantrup Fellow at Berkeley, and he has taught Anthropology at Stanford, and American Studies and Anthropology at the University of New Mexico. He is co-author of *Race, Nature and the Politics of Difference* (Duke, 2003) and author of the prize-winning *Understories: The Political Life of Forests in Northern New Mexico* (Duke, 2006), an ethnography that examines the cultural politics of nature, race and nation amid violent struggles over forests in northern New Mexico. His forthcoming book, *Homo-Apian: A Critical Natural History of the Modern Honeybee* (Duke) examines manifestations of natural history in the present, exploring contemporary taxonomies and varieties of nature, charting their resonance and discord with fossilized formations of prior natures.

Rebecca Lave is an Associate Professor in Geography at Indiana University. Her research takes a critical physical geography approach, combining political economy, STS and fluvial geomorphology to focus on the construction of scientific expertise, market-based environmental management, and water regulation. She has published in journals ranging from *Science* to *Social Studies of Science*, and is the author of *Fields and Streams: Stream Restoration, Neoliberalism, and the Future of Environmental Science* (2012, University of Georgia Press). She is co-editor of four forthcoming collections: the *Handbook of Political Economy of Science*, the *Handbook of Critical Physical Geography*, and two volumes on Doreen Massey. She also edits two book series: *Critical Environments: Nature, Science and Politics* at University of California Press (with Julie Guthman and Jake Kosek), and *Economic Transformations* at Agenda Publishing (with Brett Christophers, Jamie Peck and Marion Werner). Her current research focuses on the co-constituted hydrology, history and political economy of non-point source agricultural pollution in the US Midwest.

Tad Mutersbaugh works as a Professor of Geography at the University of Kentucky. He studies certification and assessment systems with a focus on organic agriculture in Mexico. His current research focuses upon the gender politics and political economy of biodiversity in the context of Oaxacan coffee agroforesty.

Roderick P. Neumann is Professor and Chairperson in the Department of Global and Sociocultural Studies at Florida International University. He received his PhD in Geography from the University of California in 1992 under the supervision of Michael Watts. His research explores the co-constitution of nature, society and space through a focus on modern biodiversity conservation practices. He has conducted fieldwork in Tanzania, the United Kingdom, Spain and the western United States with funding from the Social Science Research Council, Fulbright Fellowship Program, National Science Foundation and the National Endowment for the Humanities. He is the author or co-author of four books including *Imposing Wilderness* (University of California Press) and *Making Political Ecology* (Hodder-Arnold). He has also authored over 40 articles and chapters, including most recently 'Life Zones: The Rise and Decline of a Theory of the Geographic Distribution of Species' in de Bont, Raf and Jens Lachmund (eds), *Spatializing the History of Ecology: Sites, Journeys, Mappings*. New York: Routledge (in press).

Jesse Ribot is Professor of Geography, Anthropology and Natural Resources and Environmental Studies at the University of Illinois, where he is affiliated with the Unit for Criticism and Interpretive Theory and the Women and Gender in Global Perspective program, and he directs the Social Dimensions of Environmental Policy Program. Before 2008, he worked at the World Resources Institute, taught in the Urban Studies and Planning department at MIT and was a fellow at the Department of Politics of The New School for Social Research, Agrarian Studies at Yale University, the Center for the Critical Analysis of Contemporary Culture at Rutgers, Max Planck Institute for Social Anthropology, Woodrow Wilson Center and Harvard Center for Population and Development Studies. Most recently, he has been a fellow at the Stanford Center for Advanced Studies in Behavioral Sciences and an affiliate of the Department of Anthropology at Columbia University and of the Institute for Public Knowledge at New York University. Ribot is an Africanist studying local democracy, resource access and social vulnerability.

Wendy Wolford is the Robert A. and Ruth E. Polson Professor of Global Development in the Department of Development Sociology at Cornell University. She received her PhD in 2001 from Berkeley Geography and taught in the Geography Department at UNC Chapel Hill for 10 years before going to Cornell. Her work focuses on the processes, politics and promises of agrarian change.

Other Geographies, in the Work of Michael Watts

Sharad Chari, Susanne Freidberg, Jesse Ribot,
Wendy Wolford and Vinay Gidwani

*'Remembering surely stands at the center of what a critical
intellectualism must strive for'*
Michael Watts, Introduction to the 2013 edition of *Silent
Violence,* lxxxvi

Reflecting back on the 30 years since the publication of his monumental
study of food, famine and agrarian change in the Nigerian Sahel, Watts
(2013 [1983a]) notes the durability of hunger and famine, and the
importance, against the odds, of squarely facing the many fronts of
'silent violence' in the contemporary world. Over four decades of
research, writing, teaching, mentoring and performing exactly this 'critical intellectualism', Michael Watts has reconfigured our intellectual
geographies, and enabled large numbers of students, colleagues, inter-
locutors and readers of his work to further such a critical project. His
corpus has been dizzyingly wide and deep. From climate change to oil
politics, decolonization to the spectacle of the everywhere war, gendered
production politics to the commodity sensorium, African development
to documentary photography, Michael's work has pushed boundaries;
he has transformed and bridged multiple fields, including the political
economy of development, agrarian studies, political ecology, food and
famine studies, African studies, and the cultural and political economy

Other Geographies: The Influences Of Michael Watts, First Edition. Edited by Sharad Chari,
Susanne Freidberg, Vinay Gidwani, Jesse Ribot and Wendy Wolford.
© 2017 John Wiley & Sons Ltd. Published 2017 by John Wiley & Sons Ltd.

of the postcolonial South. He has written accounts of global social injustice with a close and critical attention to lived histories and geographies; and he has sought innovation in scholarly explanation, expression and advocacy. What stands out most is how prescient his work has been and how necessary it continues to be for understanding the present and the future. His generous and generative oeuvre shows how scholarly work can shape politics and praxis in multiple and unanticipated ways.

In this collection of essays, scholars touched by Michael's writing and teaching map out and discuss his influences in their work and beyond. Mirroring Michael's breadth, this collection ranges just as widely, including the political economy and ecology of African societies; governmentality and territoriality in various Southern contexts; critiques of the 'resource curse'; cultural materialist expositions of capitalism, modernity and development across the postcolonial world; extensions of the classical agrarian question in the late twentieth and early twenty-first centuries; and persisting questions of food security, hunger and famine. The collection neither intends to follow the arc of his career, nor to exhaust all his contributions to the social sciences (but see the Appendix to this introduction for a guide to his oeuvre). Rather, the authors herein discuss the efficacy of Michael's work in their own areas of research, outlining the possibilities for critical inquiry and representation that his ideas have engendered. In the spirit of engaging and extending Michael's work, each essay presents research that builds on his legacy while exploring its theoretical, analytical and empirical implications for new questions, places and times. In other words, each essay is an opening for future work to pick up a set of questions in new places, and in new ways.

Like Michael's writings, the chapters in this collection cross many categories. As a motley group of editors, we do not impose a view from on high that encompasses all possible engagements with his work; indeed this would be impossible. Neither do we aim to encompass all possible 'other geographies', following our title. There will be lacunae here for others to work with in productive ways. Rather, this introduction locates the chapters to follow in three generative areas of scholarship: political ecology, agrarian studies, and postcolonial power and praxis. These themes are intersecting rather than discreet arenas. An abiding concern with Third World and particularly African decolonization and development runs across his work and the work of several of his former students like a red thread. Similarly, Michael's oeuvre is characterized by an abiding curiosity and openness that has allowed him to think across political economy and cultural studies, anthropology and history, social science and ecology, politics and aesthetics.

Changing Frontiers of Political Ecology

It is no exaggeration to say that Michael Watts is one of the most important scholars in the development of the now well-established field of political ecology. Political ecology emerged as a critique of the field of cultural ecology in the 1970s, which was at the time the dominant framework for human-environment relations. Cultural ecology, particularly the influential approach pioneered by Andrew Vayda and Roy Rappaport (1967), analysed how individuals, households and communities adapted to particular environments, in part by regulating access to resources through complex social rules. Cultural ecologists sought to understand how what they called 'natural hazards' – droughts, famines, floods – affected these complex adaptive rule-sets, particularly in so-called Third World settings. Beginning in the late 1970s, Michael's work built on the valuable insights of cultural ecology to challenge some of the foundations of the field. He argued that cultural ecologists' Darwinian notions of adaptation naturalized a set of relationships that were in fact fundamentally shaped by power and history, and particularly by histories of colonialism and capitalism. He further argued that while hazards such as drought and other ecological or climactic conditions may be beyond an individual's control, political, social and economic conditions shaped how they are perceived and experienced – and indeed, whether or not they caused disasters.

In short, Michael showed that disaster is not natural, and that hazards are not hazards unless people are vulnerable – the precondition that transforms a natural event into a socially constructed hazard. Michael spent 15 months in Nigeria in the 1970s researching the social origins of famine among peasants. His (2013 [1983]) magnum opus, *Silent Violence: Food, Famine, and the Peasantry in Northern Nigeria*, remains one of the most important monographs in the discipline of geography today. Although journalist Alexander Cockburn, anthropologist Eric Wolf and environmental scientist Grahame Beakhurst had each previously employed the term 'political ecology', Michael's conceptual and empirical insights in *Silent Violence* helped consolidate the new field of political ecology, connecting the political economy and social history of capitalist development to human-environment relations. 'Political ecology's originality and ambition,' write Paulson, Gezon and Watts (2003, 206), 'lay in its efforts to link social and physical sciences through an explicitly theoretical approach to ecological crises that was capable of accommodating general principles and detailed local studies of problems.' Political ecology has since blossomed as a field, drawing on and developing insights from Michael's work through multiple generations of scholars, including contributors to this volume mentored by Michael. Indeed, it is through this work that the Department of Geography at UC Berkeley became known for critical, fieldwork-driven social science research on nature-society or human-environment relations.

One of Michael's first students, Judith Carney, went on to become another of political ecology's foundational scholars. In retrospect, Carney was characteristic of the emerging reconstitution of the human geography of the erstwhile colonial or developing world at Berkeley through a combination of Marxist political economy, development studies, agrarian studies, anthropology, history and ecology. Carney's early work (Carney 1986, 1988; Carney and Watts 1990, 1991) bridged the biophysical–social science divide by locating ecological inquiry within micro and macro political economies. Her chapter 'Academic Journeys in the Black Atlantic: Gender, Work and Environmental Transformations' recalls the method of intrepid curiosity that Michael has both practised and encouraged. He trained his students to pursue well-informed hunches: to situate their questions in history and place and to follow the chains of causality across time and space. In Carney's case, cultural ecology could not answer her own questions about the power hierarchies that shaped people's relations to land and other resources. In her early research on the culturally embedded, gender-differentiated effects of a Gambian rice development scheme, she turned instead to political economy, forging an approach that only later would be called political ecology. At the same time, she intervened in agrarian studies debates about the peasantry as a class, both revealing gender struggles within peasant communities and linking them to global trade policies.

While a progenitor of political ecology, Carney has also shown how the frontiers of this body of scholarship, as any other, must always shift. Indeed, Carney is a rare geographer to draw on the Black radical tradition in her important intervention on the Columbian Exchange, particularly in her 2002 book, *Black Rice: The African Origins of Rice Cultivation in the Americas*. In this and subsequent work, Carney has gone on to trace the historical connections between women's roles in rice cultivation throughout the Atlantic world – from West Africa to Brazil's *quilombos* to communities of slaves' descendants in the US South – while continuing to pull gender and race into political ecology and indeed geography more broadly, and bringing insights from what Paul Gilroy (1993) called the 'Black Atlantic' to political ecology.

As any number of textbooks, edited volumes and handbooks attest (Chambers 1989; Blaikie et al. 1994; Mearns and Norton 2010; Redclift and Grasso 2013; Kasperson, Dow and Pigeon 2016), political ecology has since broadened and diversified. It has also attracted criticism for attending more to power than ecology (Vayda and Walters 1999), and as strands of the field have moved towards post-structuralism, discourse analysis and deconstruction. Rebecca Lave has been at the forefront of a new initiative, another frontier of political ecology that aims to bridge physical geography, ecology and social science. Lave's chapter, 'Getting Back to Our Roots: Integrating Critical Physical and Social Science in the Early Work of Michael Watts', argues that the nascent field of critical physical geography (CPG) can draw lessons from early political ecology scholarship. Lave characterizes the shift to 'employ qualitative social science methods and constructivist epistemologies' as part of a 'post-structural' turn signalled by Dick Peet and Michael Watts' 1996 edited collection, *Liberation Ecologies*. While conceding that this work has yielded rich dividends in explicating the cultural politics and discursive entanglements of nature and culture (including race, gender, indigenous and class difference), her assessment is that it has increasingly neglected the material influence of biophysical processes.

As a corrective, Lave sees three key elements in the work of early political ecologists Piers Blaikie, Susannah Hecht and Michael Watts. These include first, a willingness to critically employ social and physical science, and qualitative and quantitative data, in the study of phenomena like soil erosion (Blaikie), deforestation (Hecht) and hunger (Watts). Second, Lave applauds the 'critical realism' of early political ecology, and its insistence 'that researchers could peel back … ideological layers to find true explanations'. Finally, Lave says that while both old and new political ecology accepts that nature-culture is a relation, the strength of the former is in its attention to biophysical processes. In their response to critiques of political ecology, Paulson, Gezon and Watts (2003) pose three challenges for the field: 'the first is to define politics and the environment in ways that facilitate a more thorough examination of the relationships between them; the second is to identify methods for carrying out and analysing research that encompasses relations between politics and environment; and the third is to develop ways to apply the methods and findings in addressing social-environmental concerns' (2003, 208). Lave's chapter, and the research program of CPG, deepens the question of what is to be done.

Like Watts (2013 [1983]), Lave argues for a renewed political ecology to face the immense environmental challenges of our time. As fears of climate change and environmental degradation grow, the concept of adaptation has been exhumed to walk the earth with a vengeance, not just with respect to natural hazards, but also variability, unpredictability and insecurity. From the Sustainable Development Goals to relief aid in the African horn, the drumbeat for adaptation finds renewed life through the concept of resilience (Bassett and Fogelman 2013; Watts 2015). Most major aid and development organizations now frame their work under the resilience rubric, despite a widely recognized lack of clarity about what, when or where it is. As billions of dollars are made available for those who promise to identify, quantify and promote resilience, aid agencies, foundations, non-governmental organizations and scholars have rushed to deploy the term in plans for climate adaptation, smart farming or capacity building. As Michael argues, resiliency is the social Darwinist test of one's 'right to survive' in a new global order characterized by systemic crisis and uncertainty. More than ever, Watts' (1983b) seminal article 'On the Poverty of Theory' provides a much-needed critique of adaptation, then and now, by centring labour and politics in the analysis of the resilience regime.

From the first edition of *Silent Violence* through his much later work on 'petro-violence', Michael has consistently shown how the discourses surrounding human and environmental harm can also be violent. Along with other early political ecologists, he showed how Malthusian explanations for famine and environmental degradation were not only wrong but also often alibis for wrongdoing, whether on the part of colonial regimes or postcolonial governments and donor agencies. Training his students to think dialectically, he also pushed us to question explanations that relied on binaries of any kind. Lucy Jarosz recalls this in her chapter, 'Binary Narratives of Capitalism and Climate Change: Dangers and Possibilities'. Watts' review of journalist Naomi Klein's 2015 book, *This Changes Everything: Capital versus Climate*, provides Jarosz' starting point. As the book's title suggests, Klein portrays the planet's future as a take-no-prisoners battle between capital on one hand and nature and its defenders on the other. Michael's review, while largely positive, points out that climate politics are not so neatly class-based. Some militants in the Niger Delta 'want more from oil, not less of it'. Meanwhile some corporations, especially those dependent on agricultural raw materials, have become vocal advocates of strong climate policy. Jarosz discusses how Klein's 'capital versus climate' narrative is just one of many binaries that political ecology has shown to be overly simple and analytically limited. And yet, 'they' sell books. 'They' can mobilize activism. The *New York Times* compared *This Changes Everything* to Rachel Carson's *Silent Spring*, a book that changed, if not everything, then certainly quite a lot. Jarosz concludes that while political ecologists should continue to critique binary narratives, they should also recognize their potential to inspire positive political action – and especially the type of action needed, ultimately, to achieve a world less structured around their violent simplifications.

Indeed, if Jarosz points to the persistence of dichotomous thought in rhetoric, despite our best dialectical intentions, Jake Kosek's work takes this attention to the politics of knowledge about nature in a different direction, to a different frontier of political ecology, drawing on science and technology studies (STS) and post-human or companion-species anthropology. Kosek's own intellectual trajectory, his edited work, Moore, Kosek and Pandian's (2003) *Race, Nature and the Politics of Difference*, along with Donald Moore's pioneering work on race and territory (Moore 2005), in various ways draw from Michael's scholarship on drought, famine and agrarian transformation but also his later work (1994) on the difference that 'difference' makes. In this body of work, Michael is acutely attentive to the politics of knowledge in political ecology, geography and development studies. Here, Michael pushes against positivist frameworks which pretend that objects of knowledge are readily available, innocent of the force fields within which they are made. Here, Michael owes a debt to his teacher Gunnar Ollsson, whose corpus has relentlessly worked against any kind of positivist geography. Michael's work with Allan Pred, and their (1992) book *Reworking Modernity: Capitalisms and Symbolic Discontent*, is representative of this moment in his work (also see Watts 1991), as it connects a Marxist analysis of capitalism and development to cultural politics. As he might have put it in conversation with Allan Pred or Gillian Hart at the time, their aim was to show how struggles over practice are always also struggles over meaning.

What Michael might not have anticipated are the kinds of fabrications of nature that concern Kosek, those enabled by 'big data', algorithmic innovation and quantum leaps in computing technology. In Kosek's chapter, 'Aggregate Modernities: A Critical Natural History of Contemporary Algorithms', 'the Modern Bee' is an abstraction that participates in shaping its reality. This 'statistical bee', produced by disembodying, individuating and then aggregating information from various particular honeybee colonies, has forebears in 'the taxonomic moves of the 18th and the 19th century' that 'helped produce the conditions and forms of understanding necessary to colonial projects and accumulation'. Processes of sorting human and non-human entities into classificatory systems effaced both sameness and difference, even as they manufactured new axes of association and antipathy. Taxonomies made visible previously undiscerned patterns. Moreover, by dis-embedding objects of knowledge, taxonomists laid claim to scientific authority as purveyors of the objective truth of nature. This scientific understanding was never innocent knowledge; rather, Kosek shows, it always sought mastery in order to make nature serve human ends.

The entomologist Dennis van Engelsdorp, Kosek's key protagonist, is by all accounts a talented and well-meaning scientist, who believed that if the decimation of honeybee colonies in the United States is to be arrested, the answers must come from the statistical bee. VanEngelsdorp's algorithmic alchemy is able to make 'remarkably precise' predictions of bee loss at the apiary scale, to offer beekeepers 'a perspective on their bees they had not seen before'. And yet, Kosek notes that beekeepers find it unnerving that 'a scientist who does not run bees, does not live in California, and has never seen the apiary of the beekeeper can tell that beekeeper things he himself did not know or believe about the health of his hives'. In shifting the locus of epistemic authority from beekeeper to scientist Kosek echoes Michael's recurrent concern with the tensions between the geographies of expert and experiential knowledge, whether in the realm of cultivation, soil conservation, pastureland management or water use. Like Watts, Kosek never romanticizes the local and remains circumspect about scientific knowledge that claims global relevance and becomes the touchstone for policy intervention without the active consent of those humans and non-humans whose lived environs are at stake.

Many Agrarian Questions: Peasants, Biopolitics, Enclosures

While *Silent Violence* was generative of many frontiers of political ecology, Michael's arguments in the book spoke perhaps more directly to agrarian studies, and to the revival of a much older scholarly field concerned with the transformations of agrarian society through industrialization, capitalist development, revolution and state formation. The classical 'agrarian question' posed by Marxist intellectuals at the turn of the twentieth century in fact concerned three key questions: how does capitalism 'penetrate agriculture', transforming agrarian social class structures and particularly the class of seemingly independent 'peasants'; what does agrarian industrialization, the differentiation of agrarian classes, the emergence of a sizeable agrarian proletariat and the making of a 'home market' portend for politics and in particular for the possibility of revolution; and finally how do these political and economic transformations of agrarian society, or the lack thereof, affect the broader industrial and development trajectories of regions or nations, and, by implication, of the world. The ambitions of the classical agrarian question were manifold, and it is no surprise that a strand of New Left intellectuals in the 1960s and 1970s took on these questions just as peasant movements from the Caribbean and Latin America to South East Asia were revealing themselves to be modern agents of anti-imperialism and potential socialism.

Indeed, while the student and worker protests of 'global 1968' were in many ways productive of a renewed urban studies centred on Henri Lefebvre, Manuel Castells, David Harvey and others, with lasting implications for the metropolitanism of late twentieth-century 'radical geography', these agrarian Marxists turned to other theories, and other geographies. This other strand of '1968' focused on the possibility of Third World liberation and the movement against the Vietnam War, and a renewed agrarian studies drawing on Marxist social history, Marxist feminism, development studies and anthropology. Michael's education at University College London and subsequently at the University of Michigan was shaped by these currents, in physical and/or intellectual proximity with figures like Michael Taussig and Eric Wolf. In the United States, he was also part of a moment of revival of Marxissant social history, alongside figures like Frederick Cooper and Bill Freund, and in his years in Nigeria he confronted the complexity of Marxist and radical debate over the colonial past and its postcolonial legacies. A few decades later, after he finished his PhD at Michigan and moved on to Berkeley, Michael would become an honorary member of a group of remarkable scholars, key figures in the revival of agrarian studies. This group was located in Boston and conducted empirically grounded research on the dynamics of gender, work, commodification and agrarian change in Africa and Asia, and it included Gillian Hart, Pauline Peters, Jane Guyer and Sara Berry. In this crucible of renewal, agrarian studies came together with a remarkable openness and creativity, drawing on anthropology and history, Marxism and feminism, and with a commitment to the differences constitutive of Asian, African and Latin American geographies, and Michael was very much part of it.

An important strand of debate about the agrarian question was central to the work of Eric Wolf, who taught at Michigan Anthropology during Michael's years as a graduate student. Wolf and others had engaged in a nuanced debate about the character and long-term viability of peasantries in various parts of the world. Michael's subsequent research on the persistence of peasants, and their vulnerability to famine, drew on Marxist social history and historical anthropology. Like Wolf and Sidney Mintz, Michael explored household dynamics while resolutely maintaining an understanding of the totality of social relations in Nigerian society. Michael describes the intensity of academic discourse in Nigeria at the time, with scholars in and of the region engaged in pitched debate over the nature of the nineteenth-century Sokoto Caliphate, its legacies for the nature of postcolonial social and environmental history, and particularly for the geography of Hausaland that was at the heart of *Silent Violence*. In weaving historical and theoretical threads of his analysis, Michael steered through a kind of Chayanovian populism, exemplified by Polly Hill, and a prevailing technological determinism in the time of proliferating Green Revolutions as magic fixes for food crisis in what was then called the Third World.

In the opening argument in *Silent Violence*, Watts (2013 [1983], 2–4) makes his case in relation to three threads: first, from anthropologist Keith Hart's (1982) work on the commodification of West African agriculture in an area with a historically high degree of food self-sufficiency, he notes that stronger role of the state has not meant increased agricultural productivity but rather the contrary; second, from political scientist Robert Bates' (1981) work on the dual role of the state in building a rural constituency while providing cheap urban wages, Michael extrapolates a conception of the state's role in agrarian transition as 'firmly situated in the process of capitalist development' and its varied class fractions; and, from agricultural economist Alain de Janvry's (1983) argument about 'functional dualism' in Latin American agriculture, where a dominant import-substitution industrial sector articulates uneasily with an increasingly impoverished peasant sector, with strong limits to the extent to which the latter could be squeezed, Michael draws an attentiveness to multiple modes of production in the same national social formation. While noting the long precolonial, precapitalist history of changing forms of subsistence crisis, Michael is focused on explaining the specific forms of crisis that are a consequence of the development of capitalism in northern Nigeria. This of course takes him back to history and anthropology, and to the ferment of debate in the Nigerian academy over the effects of the colonial past. The work on peasants, food, famine and capitalism remains extremely important. And as we have mentioned, his subsequent work on agrarian transformation in Senegambia brought further complexity on the relations between the politics of production and the politics of kinship, gender and conjugality. In several articles and introductions to edited collections (for example, Watts and Bassett 1985; Watts 1996; Goodman and Watts 1997), Michael has provided synoptic accounts of the multi-stranded nature of the 'agrarian question' as an analytic to understand capitalism and social change. Those he has influenced, including many of his former students, have picked up these many agrarian questions and taken them in different directions, as is evident in the following contributions to this book.

Thomas Bassett and Moussa Koné's chapter reflects the ongoing relevance of the classical agrarian studies work on peasants and capitalism. Their chapter, 'Peanuts for Cashews? Agricultural Diversification and the Limits of Adaptability in Côte d'Ivoire' offers a political-economic explanation of variations in peasant integration into markets among five cashew-producing communities. It must be said that well before the climate-change adaptation craze, Watts (2013 [1983]) critiques adaptation as a myopic concept focused on farmer adjustments and innovations. The adaptation literature had failed to recognize the constraining and damaging role of markets and the commodification of the countryside on peasant livelihoods and options. Bassett and Koné show that the Intergovernmental Panel on Climate Change (IPCC) – the premier global advising body on climate adaptation policy – portrays African peasants as lacking adaptive capacities due to an array of 'barriers' that limit their adjustment options. They characterize adaptation as incremental adjustments to these barriers. The IPCC's notion of adaptation fails to take into account the ways that peasants' options as well as what they call 'adaptive capacity' are structured by larger political-economic forces. Indeed, the inability of peasants to cope is not due to their internal 'capacity' failures; it remains an outcome of market integration.

Bassett and Koné show precisely how incomes, and the security they provide, differ among communities due to their specific place-based markets relations. Yet the IPCC continues to view adaptation as an adjustment to existing political-ecological systems – rather than a response to vulnerabilities generated by markets and government. The result is an emphasis on livelihood diversification without any attention to the commodity markets that shape the income of the poor – there is no mention of the structural and political-economic relations that shape peasant opportunity and their ability to diversify or function. There is no acknowledgement that diversification will have different meanings under different conditions: in some cases market integration can increase socioeconomic wellbeing, while in others obstacles to market access reinforce stratification and reduce crop diversity and deepen precarity. Accounting for the effects of markets requires a case-by-case empirical approach that IPCC'S blanket policy statements cannot accommodate – and so the policy recommendations tend to deepen the structural violence that reduces farmer ability to cope. Basset and Koné connect the peasant question with a call to explore a political ecology of organizations like the IPCC and the politics of their myopia (see Forsyth 2003).

While Marxist historian Eric Hobsbawm famously posed peasants as an anachronism, the content of the peasant question recurs in various places including in the North, as in the family farms of the US Midwest, or in the forms of contract farming in the US South. While peasant studies in its heyday was concerned primarily with decolonization and the Third World, there is a considerable body of work on agriculture and agrarian labour relations in the United States. Well before 'agro-food studies' changed the way we understand corporate power and agrarian commodity chains, Watts was asking questions about contract farming and commodity filières that animate some of his former students. Among those who have become key scholars in the field is Julie Guthman, whose pioneering work *Agrarian Dreams* has influenced a generation of scholars in organic or alternative agriculture. Her chapter 'Life Itself under Contract: A Biopolitics of Partnerships and Chemical

Risk in California's Strawberry Industry' revisits Watts' engagement with the peasant question as it appears in rural California. Why do 'family farms' persist even in industrial capitalist societies? Karl Kautsky first posed this question at the end of the nineteenth century, and Watts (2013 [1983]) raised it again to understand, among other things, the late twentieth-century rise of contract farming. From smallholders producing horticultural export crops in Sub-Saharan Africa to rural Americans raising broilers for corporate giants Tyson and Perdue, contract farmers offered capital the opportunity, as Guthman puts it, to 'minimize risk and maximize control'. Recognizing the broad political economic conditions that made contract relations appealing, Watts (1992, 1994a, 1994b) also attended to the 'difference that difference makes': that is, what distinguishes food from other types of commodities, and what historical, geographical and ecological conditions make specific food commodity chains themselves distinctive, and therefore worth empirical investigation. Guthman's own study of twenty-first-century California's strawberry industry builds on Peter Little and Michael Watts' (1994) edited collection, *Living Under Contract*, as well as Miriam Wells' (1996) *Strawberry Fields*. At the same time, she draws on Foucault to show how berry shippers offload not just financial risk on their so-called 'partner' growers, but also biopolitical responsibility. In other words, as berry shippers commit to phasing out the use of toxic soil fumigants, it is their growers who must find alternative ways of 'killing life to make life', which in this case means killing soil pathogens in order to harvest top quality fruit. Financially vulnerable to begin with, many of them fail at this near-impossible task. While Guthman's analysis exposes the shallowness of the industry's 'socially responsible' stance on fumigants, it also testifies to Watts' fertile contributions to agro-food studies.

Several scholars in Michael's wake connect the imperatives of agrarian studies and political ecology. The integration of agrarian studies and political ecology runs, for instance, through the work of Rod Neumann, whose first book *Imposing Wilderness: Struggles over Livelihood and Nature Preservation in Africa* (1998) is a classic in both fields. Often paired with Richard Schroeder's *Shady Practices: Agroforestry and Gender Politics in the Gambia* (1999), the two link peasant livelihoods with conservation discourses that simultaneously naturalize peasant producers and argue for their removal in order to preserve the right kind of nature. In his chapter for this book, 'Commoditization, Primitive Accumulation, and the Spaces of Biodiversity Conservation', Neumann joins Michael in insisting on the importance of history to understanding the construction of the peasantry, land use and polyvalent mechanisms of control.

In his preface to the second edition of *Silent Violence*, Michael suggests that his commitment to historical geography was nurtured by Nigerian security concerns: the military-run government 'in the white heat of Third World nationalism' (2013, lxvi) withheld permission for his field-based research for several months, giving him an excuse to focus on archival work. The deep engagement with history was necessary for developing a nuanced understanding of the relationship between capitalist development, peasant production and the changing nature of crisis. This historical perspective – in both theory and method – is well illustrated in Roderick Neumann's piece on conservation dynamics in contemporary Tanzania. Neumann draws on his past research to critique the presentist focus in the recent explosion of work on 'neoliberal conservation' and so-called 'green grabbing'. This work argues that growing interests in enclosure for the stated purpose of conserving nature constitute a 'new mode of production' whereby state expropriation fosters capital accumulation – a twisted version of the much-vaunted public private partnerships (PPP) promoted in development institutions and multilateral aid agencies. Neumann contends that a historical perspective not only provides evidence of a much longer history of enclosures for conservation ends, it also suggests that previous waves of colonial and postcolonial intervention created the very conditions of 'backwardness', 'ungovernability' and environmental collapse that provided the justification for further intervention. Reading the archives illustrates the ways in which previous economic and political movements (successive commodity booms in the Tanzanian case) re-shaped local ecologies and, in so doing, first encouraged and then dramatically undermined peasant subsistence. From ivory collection to rubber and beeswax, racialized strategies for accumulation pulled peasants into commoditized relations of production and then deprived them of the means to produce a living. As tensions in the region rose, echoing Michael's work on subsistence crises, conservation appeared as the logical political – and economic – fix. As Neumann writes, poignantly: 'Around the turn of the twentieth century, Ungindo was importing labor in the form of slaves and exporting surplus grain. Fifty years later it was exporting its labor and feeding its surplus agricultural production to wildlife.'

Concern for conservation and colonial politics continues in the focus of contemporary work by Benjamin Gardner who writes on the contradictions of Western attempts to 'save' nature while (or by) dispossessing human life in Tanzania. Ben Gardner draws on Rod Neumann's work and extends the discussion of the creation of conservation areas to newly mobile tropes of wilderness and over-population linking threats to the Serengeti national park to the destructive multiplication of Maasai peoples who are depicted as caught in the fetters of their own primitivism.

He builds on Michael's attention to the ways in which certain meanings become privileged and internalized while others are labelled alternative and reactionary. Meanings animate 'repertoires of knowledge' that become 'common sense' because they both represent and generate conditions of production and social reproduction. Gardner also builds on Michael's engagement with the spectre of Malthusianism in this process of meaning making.

In his chapter, 'Stopping the Serengeti Road: Social Media and the Discursive Politics of Conservation in Tanzania', Gardner analyses a social media campaign to 'Save the Serengeti' from a paved road the Tanzanian government proposed to build through the area. He argues that passionate arguments against the road on the Facebook page invoked the Serengeti as simultaneously unique and universal – a world heritage site essential to preservation of ecological habitat for humankind (operating as a carbon sink and home to countless endemic species) and, as such, of immense universal value. Gardner's work highlights the power relations behind knowledge production and dissemination – a Facebook campaign to 'save the Serengeti' travels considerably further and faster than arguments by 'community groups who depend on a more open understanding of conservation that includes a strong role of local people'. Underlining relations of production and power 'enabled certain repertories of knowledge to become sedimented as common sense. And … that knowledge influenced the value and meaning of material objects, places and people.' That those who argued against the road depended on an 'untouched' Serengeti for their own living was never part of the public debate highlights the dynamics of the struggle to control the circulation of meaning in and about the region.

If what is at stake in Gardner's work is a set of concerns on the enclosure of knowledge, these questions of enclosure of land and livelihoods, nature and human life extend the agrarian questions in Watts' work in new directions. Tad Mutersbaugh turns to the enclosure of knowledge, an abiding concern in his work. In his contribution to this volume, Watts' writings on oil, agrarian labour and international development provide an analytical roadmap for venturing into very different topical terrain, not exactly in agrarian studies but built on its approach. His chapter, 'Privatize Everything, Certify Everywhere: Academic Assessment and Value Transfers' combines attention to political economy, discourse and everyday work in an exploration of value transfer across the space of US higher education. As in his own writing on organic coffee, Mutersbaugh zeroes in on the certification practices that enable value to be 'piped' from one place to another. Here the valued good is not coffee or petroleum, but academic course credit, and it is moving not from (oil) field to market but rather from community colleges and other poorer institutions to 'branded' universities. Mutersbaugh shows how the creation of the national legal framework and standardized 'student learning outcomes' (SLOs) needed for credit transfers has benefited from the support of certain vested interests, including student loan agencies. He also shows why a system that would seem to benefit disadvantaged students fundamentally depends on a disadvantaged labour force, namely the adjunct faculty employed at poorer schools. His discursive analysis highlights how advocates of this system talk of the need for accountability and a 'common language', while warning of the risks of rising costs and 'curricular drift' if learning outcomes are not standardized and constantly assessed. Finally, his attention to the everyday work of assessment shows how the power relations between assessors and professors vary between institutions. Not surprisingly, the former exercise much more authority at adjunct-dominated schools than at 'branded' universities, where the tenured/tenure track faculty often resist efforts to assess and standardize their course content. Theirs is the domain that mirrors the autonomy of the peasant, but of course neither the professor nor the peasant is immune to the broader forces that relentlessly push them to conform to the dominant regime of value under capitalism. Mutersbaugh's analysis, in other words, will hit very close to home for many readers. At the same time, it testifies to just how far Watts' analytical insights can travel well beyond agrarian environs (but see Watts 1994c, 1997).

Postcolonial Power, Precarity and Struggle

Throughout his long career, Michael has maintained his connection to Nigerian political and scholarly life. Over the changing years of the late 1900s and early 2000s, he witnessed the rise of oil in national politics, and his analysis shifted to the emergence of a new state form alongside new countervailing forces of civil and uncivil society, new forms of mobilization and regional politics, as well as new extremist movements. Indeed, Michael demonstrates the power and insight born of long-term scholarly engagement with a particular national context.

Michael often characterizes his four decades of research in Nigeria as encompassing two ends of the oil boom. Indeed, one can see a thread running through his work, an attentiveness to the failures of social reproduction or to what people do when 'adaptation' is not guaranteed, whether in the devastating famines of the 1970s or in the often violent struggles for survival in the Niger Delta. Across his work, Michael has been concerned with the ways in which people living in precarious times organize and conceive of alternatives. Many of Watts' students have taken these insights in different directions, across a variety of postcolonial sites, and, in recent decades, into postcolonial urban politics and political economy.

Watts has also engaged an ambitiously high level of abstraction in his writings on what he terms 'the oil assemblage' as it structures, differentially, both the promise of untold wealth and development and the peril of emboldened power and predation. Struggles over violent accumulation in the Niger Delta also forced Watts to confront the indigenous question, in relation to his long-standing engagement with other 'Southern questions' concerning peasants, development and agrarian change. Joe Bryan's chapter, 'Oil, Indigeneity and Dispossession' picks up on this theme in at least three innovative ways. First, Bryan draws on his own long-term work with indigenous politics in the Americas. He begins with a call for solidarity at the Amazingo Institute for Indigenous Science and Technology in Ecuador, with Nigerian activists, the Ogoni 9, executed by the Abacha regime in Nigeria in 1995. Bryan grounds this act of trans-Atlantic indigenous solidarity within Watts' conception of a global and differentiated 'oil assemblage' 'whose totalizing force conjoins oil companies, state officials, security apparatuses, and local communities'. Second, however, Bryan does not rest with this mode of comparison, but rather turns to the ways in which oil, dispossession, ethnic and regional politics in different ways in these far-flung places, transform 'the very conditions of life itself'. While the category 'indigenous' proves efficacious to people in both contexts as a means of comparison, Bryan turns to the very different understandings through which these 'contrapuntal geographies' are called into being.

Bryan takes seriously the Gramscian dictum that all people on the planet are intellectuals engaged conceptually with the fundamental questions they face, through differently situated understandings of the world. Refusing to keep 'the indigenous' mired in place, in other words, Bryan asks how the category 'provides a means of thinking "contrapuntally," holding the differences in concert with one another in ways that are generative of new categories of thought and action'. Finally, Bryan juxtaposes recent Latin American work on 'extractivism' with Watts' conjunctural analysis of the oil assemblage, and to this conversation he brings his own attentiveness to concept work in the Niger Delta and Ecuador as, in his suggestive words, 'the creation of new forms of sociality fashioned to counter the forces driving dispossession'.

Indeed, Watts' work on resource extraction in relation to capital accumulation, decolonization and development mirrors the insights from a variety of places in which the struggles over the acquisition of new sources of fuel appear to generate zones of exclusion that evoke literal and metaphorical frontiers. Teo Ballvé's chapter, 'Frontiers: Remembering the Forgotten Lands', includes the perceptive insight that 'frontiers have been a loosely running thread through much of Michael Watts' work: from the silent violence of hunger in Hausaland, to the combustible oil politics of the Niger Delta, and just about everything else in between (from California communards to Boko Haram). The reason for this ... is that the agrarian question itself – one of his abiding concerns – has always been something of a frontier story.' Some of Watts' students and associates have written precisely such frontier stories, whether in Aaron Bobrow-Strain's work on the violent elite of Chiapas or James McCarthy's work on the 'wise use' movement in the US West.

Ballvé's beautifully written essay thinks productively with the 'economies of violence' that structure frontier zones, and more precisely with the region of northwest Colombia centred on Medellín. Ballvé reads Watts' attention to 'governable spaces' as deliberately ironic, as both Niger Delta and northwest Columbia conjure political disorder and violent accumulation. 'But this same messiness,' Ballvé argues, 'means frontiers can also be sites of political experimentation and possibility – the space of the rebel, the bandit, the runaway, the commoner.' Through a whirlwind of figures from agrarian elites to guerrillas, counterinsurgents to modernizing paramilitaries, Ballvé shows exactly how this political experimentation supports a process of agrarian 'laundering' of violent, elitist accumulation in the guise of communitarian and 'grassroots' development. What the paramilitaries or *paras* accomplish is, as he puts it, a Gramscian 'war of position' to counter the 'war of manoeuvre' waged against insurgents, using the latest innovations of development-speak and adapting biopolitical interventions to, as Ballvé puts it dryly, 'make live and let die; but above all, make money'. Ballvé ends with a call to 'expose over and over how these supposedly *peripheral* spaces have in fact been *central* to the creation and maintenance of our current distribution of life in the world'. His careful attention to the art of writing is also something Ballvé shares with Watts and many of his former and present students such as the talented essayist Joshua Jelly-Schapiro, or Aaron Bobrow-Strain, who is now venturing into creative non-fiction.

As we suggest, in the late 1990s, and into the 2000s, several graduate students working with Michael turned to questions concerning postcolonial urban power, capital, work and infrastructure. Some drew on the conceptual and methodological approaches of the agrarian studies literature that Michael had been a key contributor to. Others drew on his long-term engagement with the dialectics of decolonization and development, particularly on the African continent, and more particularly in the aftermath of the 'lost decade' of 1980s structural adjustment. And yet others drew on his capacious curiosity and on his attentiveness to regional differentiation and to ongoing, differentiated, material-semi-otic struggles over resources.

In a spirited afterword to a special issue of *Public Culture* on Johannesburg as a post-apartheid 'Afropolis', Watts (2006) pushes Achille Mbembe, Sarah Nuttall, AbdouMaliq Simone and other con-tributors to connect urbanism to the geography of poverty and capital. Rosalind Fredericks' 'Vibrancy of Refuse, Piety of Refusal: Infrastructures of Discard in Dakar' activates these lines of inquiry by turning specifi-cally to Michael's observation of the urbanism of 'the neoliberal tsu-nami' across African cities. Rethinking this point through Nancy Peluso's collaborative work with Watts, their (2001) edited collection *Violent Environments*, Fredericks makes the case for understanding the pro-tracted violence of neoliberal capitalism through the specific politics of garbage in Dakar. But Fredericks does much more, still, both in this piece and in her work more generally – and her restless critical engagement with multiple forces and critical traditions is also a methodological reflection of Watts' oeuvre. Fredericks provides an original response to recent work on infrastructure and materiality from ethnographic research grounded in the politics of gendered work, in the dialectics of community and citizenship, Islam and the moral politics of refusal. As we see in her piece, Fredericks engages with work by Watts on each of these themes, stretching these insights in new ways through sites of praxis from community-based waste management to 'trash revolts' and emergent claims on urban citizenship. In calling for closer attention to what she calls 'the piety of refusal', Fredericks makes space for research on the dialectics of the spiritual and material in contemporary urban social justice struggles in a variety of contexts.

Erin Collins' sweeping piece, 'Reclamation, Displacement and Resiliency in Phnom Penh' explores how land reclamation has been deployed as a tool of racialized governance in colonial, postcolonial and liberal-era Cambodia. In part, this chapter is another devastating critique of the contemporary embrace of resilience as an analytical framework for enhancing adaptive capacity under unpredictable stress. 'Resilience thinking' (Walker and Salt 2006) has been adapted from its origins in ecology and engineering, incorporating attention to social relations in part because of the strength of anti-Malthusian critiques levelled at much of the environmental literature of the 1970s and 1980s. Resilience thinking incorporates systems ecology with its focus on thresholds, tip-ping points, cycles of transformation and stability and connectivity. The flexibility of the theory contributes to its vast popularity, from the World Bank to NGOs and academics. For Watts, the rise of resilience is little more than the return of adaptation as a cultural metaphor. As he says: 'resiliency is the test of one's "right to survive" in a new global order characterized by systemic crisis, "uncertainty" and adaptation.' Collins takes on this critique of resilience in her beautifully composed and researched piece on urban land transformations. Beginning with the French in colonial Cambodia, she shows how the colonial city was carved into racialized and classed quarters, constituted through projects that turned the watery ways of the Mekong Delta into solid ground. Filling in land for the elite few, dispossessing the colonized subject and casting them on to peripheral swampland was one of the repertoires of state violence carried out on and through the water-land hybrid-scapes of Southeast Asia.

Today, Collins argues, technologies of land reclamation increasingly are implemented as a means to govern a 'climate-changed city' preparing for the worst. But the tools proffered to assist with water-related damage run aground on forms of property inculcated through previous histories of land seizure and dispossession. As Collins puts it: 'the particular imperial content of land reclamation as a technology for the making of subaltern subjects – present in recombinant form through Phnom Penh's multiple, sequential urban planning regimes – destabilizes the very "resilient" programmes and projects that work through it.' Collins incorporates Michael's attention to materiality and transformation in her argument that plans for urban development through land reclamation in Phnom Penh were never pre-determined but were shaped by the intersection of the Mekong Delta ecology, elite interests and everyday negotiation. What Collins also shows powerfully is that the political life of cities continues to be shaped by political-ecological and agrarian questions. Indeed, this essay shows how intertwined the three themes we have put forth here actually are.

In Conclusion

The essays in this volume attest to the scholarly significance of Michael's oeuvre. We note very briefly that Michael's praxis pushes out into the world beyond the academy as well, through his many years of undergraduate teaching as well as through his participation in policy communities through governmental and non-governmental organizations, and advocacy groups of various kinds. His work with photographer Ed Kashi (Watts and Kashi 2008) revives his long interest in documentary photography as a means of critique, in this case in the Niger Delta.

Michael has opened new arenas of critical inquiry that have been useful beyond the academy for framing and discerning problems concerning geography, precarity, politics and the environment. While his work has always interrogated problems of justice and wellbeing, he has sought a method always concerned with relevance to social and political change in the contexts about which we write. In a graduate seminar in the 1980s, for instance, Michael told the class to start with a problem relevant to policy and to social change, but then to completely forget about normative questions concerning change as participants conducted research. But he also insisted that they return to these normative questions at the conclusion of their research, to ask what relevance the research held for policy and practice. Whether on political ecology, agrarian questions, or on postcolonial power and struggle, Michael has always defended the relative autonomy of research, as a space of curiosity and engagement. There is no rush in this method for research to directly inform policy. Instead, there is a commitment to painstaking and long-term scholarly work. Michael has shown how 'slow research' – not surprising for a migrant transformed by many years of life in and around the alternative scenes of the San Francisco Bay Area – might engender a set of insights that slowly pries open establishment views and justifications, making space for new forms of understanding and action. In a planet of walls and prisons, and in political times that might appear bleak, we hope this praxis inspires other geographies yet to be written into the world.

References

Bassett, T. and Fogelman, C. 2013. Déjà vu or something new? The adaptation concept in the climate change literature, *Geoforum*, vol. 48, pp. 42–53.

Bates, R. 1981. *Markets and States in Tropical Africa*. Berkeley, University of California Press.

Blaikie, P. et al. 1994. *At Risk: Natural hazards, people's vulnerability, and disasters*. London, Routledge.

Carney, J. 1986. The social history of Gambian rice production, PhD thesis, Department of Geography, University of California, Berkeley.

Carney, J. 1988. Struggles over crop rights within contract farming households in a Gambian irrigated rice project, *Journal of Peasant Studies*, vol. 15, no. 3, pp. 334–349.

Carney, J. 2002. *Black Rice: The African origins of rice cultivation in the Americas*. Cambridge, MA, Harvard University Press.

Carney, J. and Watts, M. 1990. Manufacturing dissent: work, gender, and the politics of meaning in a peasant society, *Africa*, vol. 60, no. 2, pp. 207–241.

Carney, J. and Watts, M. 1991. Disciplining women? Rice, mechanization and the evolution of Mandinka gender relations in Senegambia, *Signs*, vol. 64, no. 4, pp. 651–681.

Chambers, R. 1989. Vulnerability, coping and policy (special issue Introduction), *IDS Bulletin*, vol. 20, no. 2, pp. 1–7.

De Janvry, A. 1983. *The Agrarian Question and Reformism in Latin America*. Baltimore, MD, Johns Hopkins University Press.

Forsyth, T. 2003. *Critical Political Ecology: The politics of environmental science*. New York, Routledge.

Gilroy, P. 1993. *The Black Atlantic: Modernity and double consciousness*. Cambridge, MA, Harvard University Press.

Goodman, D. and Watts, M. (eds) 1997. *Globalizing Food: Agrarian questions and global restructuring*, New York, Routledge.

Hart, K. 1982. *The Political Economy of West African Agriculture*. Cambridge, Cambridge University Press.

Kasperson, R., Dow, K. and Pidgeon, N. (eds) 2016. *Risk Conundrums: Solving unsolvable problems*. London, Earthscan.

Little, P. D. and Watts, M. 1994. Life under contract: contract farming, agrarian restructuring and flexible accumulation, in *Living Under Contract: Contract farming and agrarian transformation in Sub-Saharan Africa*. Madison, University of Wisconsin Press.

Mearns, R. and Norton, A. (eds). 2010. *Social Dimensions of Climate Change: Equity and vulnerability in a warming world*. Washington, DC, The World Bank.

Moore, D. 2005. *Suffering for Territory: Race, place, and power in Zimbabwe*. Durham, NC, Duke University Press.

Moore, D. S., Kosek, J. and Pandian, A. (eds). 2003. *Race, Nature and the Politics of Difference*. Durham, NC, Duke University Press.

Paulson, S., Gezon, L. L. and Watts, M. 2003. Locating the political in political ecology: an introduction, *Human Organization*, vol. 62, no. 3, pp. 203–217.

Peluso, N. and Watts, M. (eds). 2001. *Violent Environments*, Ithaca, NY, Cornell University Press.

Pred, A. and Watts, M. 1992. *Reworking Modernity: Capitalisms and symbolic discontent*. New Brunswick, NJ, Rutgers.

Redclift, M. R. and Grasso, M. (eds). 2013. *Handbook on Climate Change and Human Security*. Cheltenham, Edward Elgar.

Vayda, A. and Rappaport, R. 1967. Ecology, cultural and noncultural, in J. Clifton (ed.), *Introduction to Cultural Anthropology*. Boston, MA, Houghton Mifflin, pp. 457–477.

Vayda, A. and Walters, B. 1999. Against political ecology, *Human Ecology*, vol. 27, pp. 167–179.

Walker, B. and Salt, D. 2006. *Resilience Thinking*. Washington, DC, Island Press.

Watts, M. 1983a. *Silent Violence: Food, famine, and peasantry in Northern Nigeria*. Berkeley, University of California Press.

Watts, M. 1983b, On the poverty of theory: natural hazards research in context, in K. Hewitt (ed.), *Interpretations of Calamity from the Viewpoint of Human Ecology*. Boston, MA, Allen & Unwin.

Watts, M. 1991. Mapping meaning, denoting difference, imagining identity, *Geografiska Annaler, Series B, Human Geography*, vol. 73, no. 1, pp. 7–16.

Watts, M. 1992. Peasants and flexible accumulation in the third world: producing under contract, *Economic and Political Weekly*, vol. 27, no. 30, pp. PE90–PE97.

Watts, M. 1994a. What difference does difference make?, *Review of International Political Economy*, vol. 1, no. 3, pp. 563–570.

Watts, M. 1994b. Life under contract: contract farming, agrarian restructuring and flexible accumulation, in P. D. Little and M. J. Watts (eds), *Living Under Contract: Contract farming and agrarian transformation in Sub-Saharan Africa*, Madison, University of Wisconsin Press, pp. 21–77.

Watts, M. 1994c. Development II: the privatization of everything, *Progress in Human Geography*, vol. 18, no. 3, pp. 371–384.

Watts, M. 1996a. Development III: The global agro-food system and late twentieth century development (or Kautsky redux), *Progress in Human Geography*, vol. 20, no. 2, pp. 230–245.

Watts, M. 1996b. *Liberation Ecologies: Environment, development and social movements* (co-edited with Richard Peet). New York, Routledge.

Watts, M. 1997. African studies at the fin de siècle: is it really the fin? *Africa Today*, vol. 44, no. 2, pp. 185–191.

Watts, M. 2006 Baudelaire over Berea, Simmel over Sandton, *Public Culture*, vol. 17, no. 1, pp. 181–192.

Watts, M. 2013. [1983a] *Silent Violence: Food, famine, and peasantry in Northern Nigeria*. Athens, University of Georgia Press.

Watts, M. 2015. Now and then: the origins of political economy and the rebirth of adaptation as a form of thought, in T. Perreault, G. Bridge and J. McCarthy (eds), *The Routledge Handbook of Political Ecology*. Abingdon, UK, Routledge, pp. 19–50.

Watts, M. and Bassett, T. J. 1985. Crisis and change in African agriculture: a comparative study of Ivory Coast and Nigeria, *African Studies Review*, vol. 28, no. 4, pp. 3–27.

Watts, M. 2008. *Curse of the Black Gold: 50 years of oil in the Niger Delta* (with Ed Kashi). Brooklyn, NY, powerHouse Books.

Wells, M. 1996. *Strawberry Fields: Politics, class and work in California agriculture*, Ithaca, NY, Cornell University Press.

Appendix: The Arc of Michael Watts' Thought

1977. Modernization and Social Protest Movements (with Arthur Drysdale), *Antipode*, vol. 9, no. 1, pp. 40–55.

1983. *Silent Violence: Food, famine, and peasantry in Northern Nigeria*. Berkeley, University of California Press.

1983. Hazards and crises: a political economy of drought and famine in Northern Nigeria, *Antipode*, vol. 15, no. 1, pp. 24–34.

1983. On the poverty of theory: natural hazards research in context, in K. Hewitt (ed.), *Interpretations of Calamity from the Viewpoint of Human Ecology*. Boston, MA, Allen & Unwin.

1983. The political economy of climatic hazards: a village perspective on drought and peasant economy in a semi-arid region of West Africa, *Cahiers d'Etudes Africaines*, vol. 23, no. 89/90, pp. 37–72.

1985. Social theory and environmental degradation: the case of Sudano-Sahelian West Africa, in Y. Gradus (ed.), *Desert Development: Man and technology in sparselands*. Dordrecht, Netherlands, Reidel.

1985. Crisis and change in African agriculture: a comparative study of Ivory Coast and Nigeria (with Tom Bassett), *African Studies Review*, vol. 28, no. 4, pp. 3–27.

1986. Geographers among the peasants: power, politics and practice, *Economic Geography*, vol. 62, no. 4, pp. 373–386.

1988. The faces of famine: a response to Torry, *GeoJournal*, vol. 17, no. 1, pp. 145–149.

1989. The agrarian question in Africa: debating the crisis, *Progress in Human Geography*, vol. 13, no. 1, pp. 1–41

1990. Peasants under contract: agro-food complexes in the Third World, in H. Bernstein, B. Crow, M. Mackintosh and C. Martin (eds), *The Food Question*. London, Monthly Review Press, pp. 149–162.

1990. Manufacturing dissent: work, gender, and the politics of meaning in a peasant society (with Judith Carney), *Africa*, vol. 60, no. 2, pp. 207–241.

1991. Disciplining women? rice, mechanization, and the evolution of Mandinka Gender relations in Senegambia, *Signs*, vol. 16, no. 4, pp. 651–681.

1991. Mapping meaning, denoting difference, imagining identity, *Geografiska Annaler*, Series B, *Human Geography*, vol. 73, no. 1, pp. 7–16.

1991. Entitlements or empowerment? famine and starvation in Africa, *Review of African Political Economy*, vol. 51, pp. 9–26.

1992. Peasants and flexible accumulation in the Third World: producing under contract, *Economic and Political Weekly*, vol. 27, no. 30, pp. PE90–PE97.

1992. *Reworking Modernity: Capitalisms and symbolic discontent* (with Allan Pred). New Brunswick, Canada, Rutgers University Press.

1992. Space for everything (A Commentary), *Cultural Anthropology*, vol. 7, no.1, pp. 115–129.

1993. The space of vulnerability (with Hans Bohle), *Progress in Human Geography*, vol. 17, no. 1, pp. 43–67.

1993. Hunger, famine and the space of vulnerability (with Hans Bohle), *GeoJournal*, vol. 30, no. 2, pp. 117–125.

1993. Development I: power, knowledge, discursive practice, *Progress in Human Geography*, vol. 17, no. 2, pp. 257–272.

1994. Life under contract: contract farming, agrarian restructuring and flexible accumulation, in P. D. Little and M. J. Watts (eds), *Living Under Contract: Contract farming and agrarian transformation in Sub-Saharan Africa*, Madison, University of Wisconsin Press, pp. 21–77.

1994. What difference does difference make?, *Review of International Political Economy*, vol. 1, no. 3, pp. 563–570.

1994. Development II: the privatization of everything, *Progress in Human Geography*, vol. 18, no. 3, pp. 371–384.

1995a. Working-class heroes, *Transition*, vol. 68, pp. 90–115.

1995b. *Geographies of Global Change: Remapping the world in the late twentieth century* (co-edited with P. Taylor and R. J. Johnston). Oxford, UK, Basil Blackwell.

1996. Islamic modernities: citizenship, civil society, and Islamism in a Nigerian city, *Public Culture*, vol. 8, pp. 251–289.

1996. Development III: the global agro-food system and late twentieth century development (or Kautsky Redux), *Progress in Human Geography*, vol. 20, no. 2, pp. 230–245.

1996. *Liberation Ecologies: Environment, development and social movements* (co-edited with Richard Peet). New York, Routledge.

1997. African Studies at the Fin de Siècle: Is it Really the Fin?, *Africa Today* 44(2, pp. 185-191.

1997. *Globalizing Food: Agrarian Questions and Global Restructuring* (co-edited with David Goodman). London, Routledge.

1998. Nature as artifice and artifact. In B. Braun and N. Castree (eds), *Remaking Reality*. London, Routledge, pp. 243–268.

1999. Privatization and governance, *Ethics, Policy and Environment*, vol. 2, pp. 256–257.

2000. Political ecology, in E. Sheppard and T. Barnes (eds), *A Companion to Economic Geography*. Oxford, UK, Blackwell, pp. 257–274.

2000. Development at the millennium: Malthus, Marx and the politics of alternatives, *Geographische Zeitschrift*, vol. 88, no. 2, pp. 67–93.

2000. The Great Tablecloth: bread and butter politics, and the political economy of food and poverty, in G. Clark, M. Gertler and M. Feldmann (eds), *A Handbook of Economic Geography*. London, Oxford University Press, pp. 195–215.

2000. Contested communities, malignant markets, and gilded governance, in C. Zerner (ed.), *People, Plants and Justice*. New York, Columbia University Press, pp. 21–51.

2000. Poverty and the politics of alternatives at the end of the millennium, in J. Nederveen Pieterse (ed.), *Global Futures*. London, Zed Press, pp. 133–147.

2001. *Violent Environments* (co-edited with Nancy Lee Peluso), Ithaca, NY, Cornell University Press.

2001. 1968 and All That..., *Progress in Human Geography*, vol. 25, no. 2, pp. 157–188.

2001. Violent geographies: speaking and unspeakable and the politics of space, *City and Society*, vol. XIII, no. 1, pp. 85–117.

2001. Development ethnographies, *Ethnography*, vol. 2, no. 2, pp. 283–300.

2001. Black acts, *New Left Review*, vol. 9, pp. 125–139.

2002. Chronicle of a death foretold: some thoughts on peasants and the agrarian question, *Oesterreichische Zeitschrift fuer Geschichtswissenschaften*, vol. 4, pp. 22–51 (and Commentary pp. 51–61).

2002. Hour of darkness, *Geographica Helvetica*, vol. 57, no. 1, pp. 5–18.

2002. Preface: green capitalism, green governmentality, *American Behavioral Scientist* (special issue), vol. 45, no. 9, pp. 1313–1317.

2003. Locating the political in political ecology: an introduction (special issue, with co-authors), *Human Organization*, vol. 62, no. 3, pp. 205–217.

2003. Alternative modern: development as cultural geography, in S. Pile, N. Thrift, K. Anderson and M. Domosh (eds), *Handbook of Cultural Geography*. London, Sage, pp. 433–453.

2003. Economies of violence: more oil, more blood, *Economic and Political Weekly*, vol. 38, no. 48, pp. 5089–5099.

2003. Development and governmentality, *Singapore Journal of Tropical Geography*, vol. 24, no. 1, pp. 6–34.

2003. Thinking with the blood: a response to Reginald Cline-Cole, J. K. Gibson-Graham and Marcus Power, *Singapore Journal of Tropical Geography*, vol. 24, no. 2, pp. 258–262.

2004. Antinomies of community, *Transactions of the Institute of British Geographers, New Series*, vol. 29, no. 2, pp. 195–216.

2004. Resource curse? Governmentality, oil and power in the Niger Delta, *Geopolitics*, vol. 9, no. 1, pp. 50–80.

2005. Righteous oil? Human rights, the oil complex, and corporate social responsibility, *Annual Review of Environmental Resources*, vol. 30, pp. 373–407.

2005. Nature/culture: a natural history, in R. Johnston and P. Cloke (eds), *Spaces of Geographical Thought*. London, Sage, pp. 142–174.

2005. Scarcity, modernity, terror, in B. Hartmann, B. Subramanian and C. Zerner (eds), *Making Threats: Biofears and environmental anxieties*. New York, Rowman and Littlefield, pp. 99–106.

2005. What's left? Left retort, *Antipode*, vol. 37, no. 4, pp. 643–653.

2006. Baudelaire over Berea, Simmel over Sandton, *Public Culture*, vol. 17, no. 1, pp. 181–192.

2006. Culture, development and global neoliberalism, in S. Radcliffe (ed.), *Culture and Development in a Globalising World*. London, Routledge, pp. 30–58.

2006. In search of the Holy Grail: projects, proposals and research design, in E. Perecman (ed.), *Method is the Madness*. New York, Sage, pp. 175–197.

2006. Empire of oil, *Monthly Review* (September), pp. 1–17.

2007. Petro-insurgency or criminal syndicate? Conflict & violence in the Niger Delta, *Review of African Political Economy*, vol. 34, no. 114, pp. 637–660.

2007. What might resistance to neoliberalism consist of?, in N. Heynen, J. McCarthy, S. Prudham and P. Robbins (eds), *Neoliberal Environments: False promises and unnatural consequences*. New York, Routledge, pp. 273–278.

2007. Revolutionary Islam: a geography of modern terror, in D. Gregory and A. Pred (eds), *Violent Geographies*. London, Routledge, pp. 170–203.

2008. Imperial oil: the anatomy of a Nigerian oil insurgency, *Erdkunde*, vol. 62, no. 1, pp. 27–39.

2008. *Curse of the Black Gold: 50 years of oil in the Niger Delta* (with Ed Kashi). Brooklyn, NY, powerHouse Books.

2008. Economies of violence: more blood, more oil, in A. Baviskar (ed.), *Contested Grounds: Essays on nature, culture and power*. New Delhi, Oxford University Press, pp. 106–136.

2008. Anatomy of an oil insurgency, in K. Omeje (ed.), *Extractive Economies and Conflicts in the Global South*. London, Ashgate, pp. 51–74.

2008. The price of our oil addiction, in D. Elliot Cohen (ed.), *What Matters: The world's preeminent photojournalists and thinkers depict essential issues of our time*. New York, Sterling Publishers, pp. 222–242.

2008. Petroleum in Africa, *Encyclopedia of the Modern World, Vol. 3*. New York, Oxford University Press.

2008. Soft machine: notes on oil addiction, *Human Geography*, vol. 1, no. 2, pp. 33–41.

2009. The rule of oil: petro-politics and the anatomy of an insurgency, *Journal of African Development*, vol. 2, pp. 27–56.

2009. Now and then, *Antipode*, vol. 41, no. 6, pp. 1152–1168 (reprinted in N. Castree et al., (eds), *The Point is to Change It*. Oxford, UK, Blackwell, pp. 10–26).

2009. Radicalism, writ large and small, in J. Pugh (ed.), *What Is Radical Politics Today?* London, Palgrave, pp. 103–112.

2009. Reflections, Michael Watts interviewed by Murat Arsel, *Development and Change, vol. 40, no. 6, pp. 1191–1214*.

2009. The Southern Question, in H. Akram-Lodhi and C. Kay (eds), *Peasants and Globalization: Political economy, rural transformation and the agrarian question*. Abingdon, UK, Routledge, pp. 262–287.

2010. *Global Political Ecology* (co-edited with Richard Peet and Paul Robbins). New York, Routledge.

2010. Oil city: petro-landscapes and sustainable future, in G. Doherty and M. Mostafavi (eds), *Ecological Urbanism*. Baden, Germany, Lars Muller Publishers, pp. 420–430.

2000/2005/2009/2011. *Dictionary of Human Geography* (co-edited with Ron Johnston, Gerry Pratt and Derek Gregory). London, Wiley-Blackwell,

2011. On confluences and divergences, *Dialogues in Human Geography*, vol. 1, no. 1, pp. 84–89.

2011. Ville petrolière: petro-paysages et futurs soutenables, *Ecologie et Politique*, vol. 42, pp. 65–71.

2011. Petroland: Schaften Unde die Nachhaltigkeit der Zukunft, in *The Oil Show*, curated by Inke Arns, Hartware MeienKunstVerein (Dortmund University). Berlin, Verlag (Revolver Publishing), pp. 39–51.

2011. Blood oil, in A. Behrends, Stephen P. Reyna and G. Schlee (eds), *Crude Domination: An anthropology of oil*. Oxford, UK, Berghahn Books, pp. 49–80.

2011. Crimini dimenticati: vita, morte e inganno nei giacimenti petroliferi della Nigeria (Forgotten crimes: life, death and illusion on the Nigerian oilfields), 900, *La Fine del Petrolio*, no. 4, pp. 133–166.

2011. Turbulent oil: conflict and insecurity in the Niger Delta (with Ibaba Samuel Ibaba), *African Security*, vol. 4, no. 1, pp. 1–19.

2011. Planet of the wageless, *Identities*, vol. 18, no. 1, pp. 69–80.

2011. Ecologies of Rule: African Environments and the Climate of Neoliberalism, in C. Calhoun and G. Derlugian (eds), *The Deepening Crisis: Governance challenges after neoliberalism*, Possible Futures Series, vol. 2. New York, New York University Press, pp. 67–92.

2012. *West of Eden: Communes and utopia in Northern California* (co-edited with Iain Boal, Cal Winslow and Janferie Stone). Oakland, CA, PM Press.

2012. Caught on the hop of history: communes and communards on the canvas of 1968, in Iain Boal et al. (eds), *West of Eden: Communes and utopia in Northern California*, pp. 249–271.

2012. Sweet and sour: the oil curse in the Niger Delta, in T. Butler, D. Lerch and G. Wuerthner (eds), *The Energy Reader*. Santa Rosa, CA, Post Carbon Institute, pp. 247–256.

2012. A tale of two gulfs: life, death and dispossession along two oil frontiers, *American Quarterly*, vol. 64, no. 3, pp. 437–467 (reprinted in P. Chakrabarty and D. da Silva (eds), *Race, Empire and the Crisis of the Subprime*. Baltimore, MD, Johns Hopkins University Press, 2013).

2012. Globophobia, in G. Ritzer (ed.), *Wiley-Blackwell Encyclopedia of Globalization*. London, Blackwell.

2012. Modular development, diminutive capitalists and the financialization of capitalism, *Antipode*, vol. 44, no. 2, pp. 535–541.

2012. Of bats, birds and mice, in C. Abrahamson and M. Gren (eds), *Gunnar Olsson: On the Geographies of Gunnar Olsson*. Aldershot, UK, Ashgate, pp. 143–155.

2012. Nigeria and the politics of price, *The Mark*, 25 January.

2012. Economies of violence: a critique of the World Development Report 2011, *Humanity*, vol. 3, no. 1 (Spring), pp. 115–130.

2013. Bare life and the new interregnum, new Introduction to the 2nd edition of *Silent Violence: Food, famine and peasantry in Northern Nigeria*. Athens, University of Georgia Press.

2013. Insurgent geographies (in German), *Geographisches Rundschau*, vol. 9, pp. 44–52.

2013. A tale of two insurgencies: oil, authority and the spectre of terror in Nigeria, in Alex Houen (ed.), *States of War in the War on Terror*. Cambridge, UK, Polity Press, pp. 103–129.

2013. Oil talk, *Development and Change*, vol. 44, no. 4, pp. 1013–1026.

2013. Imperiales Oil und vergessene Verbrechen: Grenzgebiete der Enteignung in Niger-Delta (Imperial Oil and Forgotten Crimes: Frontiers of Dispossession in the Niger Delta, Nigeria), *PROKLA*, vol. 43, no. 1, pp. 61–71.

2013. Toward a political ecology of environmental security, in R. Floyd and R. Matthew (eds), *Environmental Security*. London, Routledge, pp. 82–102.

2014. Resource violence (with Nancy Lee Peluso), in Carl Death (ed.), *Critical Environmental Politics*. London, Routledge, pp. 184–197.

2015. *Oil Talk: The secret lives of the oil and gas industry* (co-edited with Arthur Mason and Hannah Appel). Ithaca, NY, Cornell University Press.

2015. Chronicle of a future foretold: the complex legacies of Ken Saro-Wiwa, *The Extractive Industries and Society*, vol. 2 (August), pp. 635–664

2015. Now and then: the origins of political economy and the rebirth of adaptation as a form of thought, in T. Perreault, G. Bridge and J. McCarthy (eds), *The Routledge Handbook of Political Ecology*. Abingdon, UK, Routledge, pp. 19–50.

2016. *The mafia of a Sicilian village, 1860–1960; a study of violent peasant entrepreneurs*, by Anton Blok, Agrarian Classics Review Series, *The Journal of Peasant Studies*, vol. 43, no. 1, pp. 67–91.

2016. The precarious lives of the commons: voices and lessons from the oilfields of the Niger Delta, in C. Winslow (ed.), *River of Fire: Commons, Crisis and Imagination*. Arlington MA, The Pumping Station.

2017. Joined at the head: anthropology, geography and the environment, in S. Coleman, S. B. Hyatt and A. Kingsolver (eds), *The Routledge Companion to Contemporary Anthropology*. Abingdon, UK, Routledge, pp. 323–358.

1

Academic Journeys in the Black Atlantic

Gender, Work and Environmental Transformations

Judith Carney

The University of California at Berkeley during the mid-1970s was an exhilarating place and time for graduate students in the Department of Geography. As a group we were much affected by the pervasive social inequalities and environmental transformations that were radically changing the world. The course curriculum in the department was unstructured but encouraged engagement with the University's natural science programmes. We studied botany, forest ecology, soils, rangeland management, pest management and non-Western food production systems; we took courses in anthropology, history and agricultural economics. This multidisciplinary approach, of course, is a hallmark of Geography, and its intention to broker and synergize the biophysical–social science divide is one of the discipline's great virtues. Thus, inspired by our Berkeley training, with its strong emphasis on fieldwork and language acquisition, an invigorated cohort of graduate students began their scholarly adventures.

Into this highly energized scene alighted Dr Michael Watts, a newly minted PhD who arrived on campus in 1979 as assistant professor of Geography. To this day, I remember his electrifying job talk on hunger, poverty and famine in northern Nigeria. I realized immediately that Michael was asking precisely the questions I wanted to work on in my own research. Indeed, what drew me to Geography, graduate school,

Other Geographies: The Influences Of Michael Watts, First Edition. Edited by Sharad Chari, Susanne Freidberg, Vinay Gidwani, Jesse Ribot and Wendy Wolford.

and specifically Berkeley, was a book I read while working in Oregon: *Agricultural Origins and Dispersals* by Berkeley geographer Carl O. Sauer. I found Sauer's holistic perspective on culture and the environment engaging and compelling. His work gave me an appreciation for a discipline that placed human-environmental relations at the forefront of analysis at a time in the 1970s when the environmental movement was gathering momentum. Sauer's writings on cultural landscapes, indigenous peoples' contributions to crop and animal domestication, and Latin America inspired me during my first travels through Mexico and Central America. When I arrived at Berkeley, I learned that Professor Sauer had long retired from the faculty; however, there were a core group of faculty and graduate students who shared my enthusiasms for Latin America, environmental change, human ecology and poverty.[1] Michael's arrival added an enlivened coda to this group. He introduced us to the theoretical literatures on the structural forces within the world system that produce wealth and poverty and pointed to how these ideas might cohere with and clarify our own research. I felt Michael's work added a critical perspective to the inspiration I took from Sauer's work and the 'Berkeley school's' emphasis on cultural ecology. There was a sense that an older generation of geographers was ceding the stage to a new and exciting one.

I had begun my PhD research in Berkeley – before Michael's arrival – on South America, specifically, Amazonic Brazil. By 1977 dramatic landscape changes and population movements were accompanying the military government's programme to develop the Amazon for mineral extraction, cattle ranching, tree farms and pulp mills. Road construction and sponsored colonization schemes attracted new settlers into the region, contributing to alarming trends in deforestation. Such socio-economic and environmental changes were already proceeding apace in rural Maranhão, where I was headed. Roads were reaching the state's forested interior and the peoples who lived there. Land expropriations, ineffective World Bank projects, mineral companies and logging interests soon followed, undermining the way of life of indigenous peoples and the mixed-race farmers (*caboclos*), who traditionally planted rice and tropical tubers for subsistence and harvested *babaçu* palm forest products for food and trade. Their territory was now a new frontier for outside interests, who were ready and willing to dispossess smallholder farmers by force, if necessary, from mostly untitled land. But I was unaware of all this until a 16-hour bus ride deposited me in the Alto Turiaçu region, where I immediately felt the unease born of conflict. Instead of finding myself in an area amenable to my academic project – a cultural ecological analysis of smallholder environmental adaptation and resilience – I found a region in turmoil. There, I met university students wanted by the military government for their political activities, communist leaders of peasant leagues, activist Catholic priests, gun-slinging enforcers for cattlemen, and FUNAI government agents charged with 'assimilating' the indigenous Ka'apor people.[2] Nothing in my academic training had prepared me for what I experienced during those weeks in the interior of Maranhão. There, too, I first encountered *quilombos*, free communities that runaway slaves founded during slavery still inhabited by their descendants.[3]

Throughout the Alto-Turiaçu region, smallholder cultural identity and memory was rooted in an environment that was undergoing unprecedented and often violent transformation. What had captured my attention during this visit to rural Maranhão was the disquieting struggle of the rural poor to stay on the land. As neither *quilombolas* nor peasant farmers possessed formal title to the land their families had worked for generations, they had no legal standing to defend their claims.[4] For the military government and its economic allies, they were impediments to the process of transforming the Amazon. The new invaders frequently used guns and intimidation to seize land continuously held by the mixed-race descendants of runaway slaves. Many rural Maranhense saw their communities destroyed as systemic violence forced them to flee. Some became part of the huge exodus of poor Brazilians who crowded into colonization projects that frequently failed them.

My experience in turbulent rural Maranhão in 1977 prompted two critical insights that have since informed much of my research. First, it provided a way of seeing the world – the interplay of culture and environment, the relationship of place-based knowledge systems to subsistence strategies and environmental sustainability. I began to see the limits of the cultural ecology research framework in which I was being trained since it failed to provide the tools to examine land use in the context of economic change and policies that privileged one type of land use over another. The economic interests that freely exercised frontier justice to force peasant farmers off their land were threatening an entire way of life whose roots were Amerindian and New World African. Having witnessed the shooting of one man for resisting efforts to seize his land, I was no longer satisfied with academic discussions that presumed stability in local knowledge and land use.

Like many other geographers working in the tropics where military governments ruled, I turned to political economy and to the development studies literature for a more dynamic conceptualization of what was occurring. In this way, the accomplishments and struggles of rural peoples could be placed against the background of economic change and power relations. I was among the first generation of geographers whose fieldwork in the global South demanded a more robust theoretical framework – one that would later become known as political ecology.[5] Michael Watts played no small part in the development of this framework.

There was a second insight from my fieldwork in Maranhão that I did not fully appreciate at the time. My work among the *quilombos* had actually provided a powerful introduction to the African presence in tropical America. When I subsequently went to West Africa for my dissertation research, I began to think about African continuities, the cultural heritage of enslaved Africans in Neotropical environments, and the creolized food systems they shaped in former plantation societies. The fieldwork experience on both sides of the Atlantic moved me to think about the Atlantic Basin as a historical-geographical continuum. When nearly 25 years later I returned to rural Maranhão, it was with the realization that many of the subsistence farming practices that I witnessed in the 1970s incarnated profound connections to West Africa.

But very little of this was clear when I first met Michael Watts. The urgent questions he was asking about Nigeria and Africa ran congruent to my own rather inchoate thoughts, especially as they turned to my experiences in Brazil. Unhappily, I realized that it was not possible to explore these questions in Maranhão because my dissertation chair was not supportive. I passed my PhD exams to work in Brazil but found the pull of questions difficult to ignore. It was then, with Michael's help and sponsorship, I managed the awkward transition to a new committee chair and study region. As a junior assistant professor, Michael had now assumed mentorship of a would-be Africanist who had advanced to candidacy but never taken a course on Africa. Weighed down by these handicaps, I was unable to qualify for a competitive fellowship. Fortunately, Michael intervened at a critical moment: as co-director of a study on the Gambia and Senegal River basins sponsored by the Center for Research on Economic Development (CRED) at the University of Michigan, he was able to enlist me in the CRED project, thereby enabling me to go to West Africa for the first time in 1983. Michael was the reason why I so fortuitously ended up in The Gambia rather than Brazil for my dissertation.

West African rivers had become the focus of international financial aid in the decade following the 1968–73 Sahel Drought. The field director of the Gambia River Basin Studies was the noted ecologist, Karl F. Lagler. Lagler believed firmly in scholarship that combined biophysical with social science research. We agreed that I would examine the downstream socio-economic impact assessment of a proposed anti-salinity barrage along the Gambia River. The Gambia was a rice-growing country. Development planners sought to 'drought proof' the region by converting wetlands to pump-irrigated rice schemes that would support the production of two annual rice crops. The proposed development promised the rural poor both food security and cash income from the sale of surplus crops. But, as I discovered during my study, rice is traditionally a woman's crop in The Gambia and the proposed drought-proofing project would have a direct and significant impact on their daily lives, work and income. Government officials and international donors proclaimed the project was a model for gender-sensitive development; it would deliver direct benefits to women farmers. Suspecting that I might be witnessing an epochal agrarian transition, I decided to focus on the 'gender' question of rice production for my dissertation. I relocated to an agricultural research station in rural Gambia to begin my fieldwork.

Gender and Agrarian Transitions

I began my 14-month field research with a survey of farming practices in the central region of the country. These encompassed wetlands of the Gambia River, the focus of rice development projects and drier uplands where peanut cultivation prevailed. The wetlands were the principal zone of food production during the colonial period, while the rain-fed plateau specialized in peanut cultivation for export. Rice was a subsistence crop cultivated by women, while men cultivated peanuts as a cash crop. Colonial policies had encouraged a greater reliance on women's swamp rice production for food, but this was never enough and was augmented by rice imports. Since the end of colonial rule, the country had undergone an incomplete process of agrarian transformation. Commodity production of peanuts had reduced the rainfed upland acreage devoted to subsistence cereals (millet, sorghum, maize), but cash income from peanuts, as with African commodities in general, had declined in global markets. Droughts had further strained subsistence production of rice, and the country relied on imports for around half its rice consumption. It was in this context that irrigated rice projects promised to make up the domestic shortfall by extending cash cropping to wetland environments. This meant that a subsistence crop traditionally grown by women was going to be commodified.

The development studies literature Watts included in his seminars drew attention to the role of the state, markets and class for understanding the power relations mediating specific agrarian transitions such as the Gambian one I witnessed in the 1980s. His seminars also introduced students to debates in peasant studies, such as whether specific peasantries were differentiating with market development, the moral economy that at times enabled family labour to intensify, peasant resistance to labour exploitation, and peasant-based political movements. Amartya Sen's pioneering scholarship linking hunger and famine to policies that weakened the claims of the poor to subsistence entitlements proved another important influence.[6]

However, when I left for West Africa to start my fieldwork, the literature on agrarian political economy, peasant and development studies, and cultural ecology in Geography had little to say about rural women, households or gender relations. The debates on markets, states and the peasantry as a class inevitably overlooked the importance of women in subsistence agriculture. Some female social scientists (principally in rural sociology and development economics and anthropology), were beginning to bring women into the discussion but mainly through the analytical framework of class.[7] There was little geographical scholarship on agrarian transformations and their impact on women's labour, perhaps because not many female geographers were engaged at that time in international development fieldwork.

Gambian farming systems had been studied since the 1940s by a few excellent scholars. Their work provided an invaluable historical perspective on the role of rice farming in subsistence strategies from the colonial era.[8] While in the field, I was not yet aware of the household studies literature that was emerging as a critique of Marxist class analysis of agrarian transformations. Watts introduced me to this literature upon my return to Berkeley in 1985. Feminist concern with reproduction and production had opened up the household as a category of analysis, especially for examining demands on female labour, income opportunities, the social construction of gender, and patriarchy within and between households. Household studies revealed that the peasantry as a class was shaped by many types of responses to poverty. The burdens placed upon women and differential access to and control over resources were significant for understanding how peasant households responded to agrarian change and the effect of economic transformation on patriarchal family structures.[9]

Returning from my fieldwork in central Gambia, I immediately recognized the salience of this scholarship for understanding the gender conflicts that ensued in the Jahaly-Pacharr irrigated rice scheme. Jahaly-Pacharr was promoted as a 'women's project' but the Gambian government officials registered the irrigated plot allotments to male heads of households. By registering titles in men's names, project officials claimed that it would break down a gendered cropping system. The project's work calendar was calibrated on the availability of both male and female family labour for year-round rice cultivation despite longstanding male resistance to working on a 'woman's crop'. Tension over which household members would carry the household labour burden in irrigated rice and who would be the beneficiaries of paddy sales became evident after the first harvest, when the male household head listed on the land registry received the payment. The wives and daughters had laboured to produce the crop in a project which formally enabled senior male members to control the earnings. Thus technological change in rice cultivation, which created conditions for two cropping seasons and surplus production, had effectively intensified women's labour while transferring their traditional control of plots and income benefits to their male household heads.

When women were denied the benefits from sales of the first harvest, there followed a loud and unprecedented outcry during the second planting season. They refused further work unless compensated in paddy. Some households conceded to their demands; others did not. Harvest yields declined as the project's planting calendar depended upon the full, yet unrealized, participation of family labour, not just females. After completion of my dissertation in 1986, Michael and I incorporated these insights into two well-received research articles that addressed the crucial role of family authority relations and property relations for structuring gender divisions of labour and access to and control over rural resources in agrarian transitions of the global South.[10]

By the mid-1990s policy reforms had delivered the final blow to the Jahaly-Pacharr scheme. IMF structural adjustment policies had removed the price support for domestically grown rice in favour of cheaper imports. Economic reforms and donor assistance no longer prioritized 'food security' as a development strategy, favouring instead 'comparative advantage' and market liberalization. The policy shift, which eliminated protective tariffs for domestic rice growers just as fertilizer prices quadrupled in The Gambia, dashed the country's hopes for rice self-sufficiency.

Despite proclamations of "food security' and 'a women's project', irrigated rice development had not improved women's socio-economic position. But I saw a way forward through advocacy of traditional swamp rice growing as an alternative, 'bottom-up' strategy that would positively reach female growers. I devoted one research trip to deepening my understanding of the agro-ecological practices and micro-environments in which women customarily planted rice, that I had initiated with the CRED study. The research represented a return to earlier graduate-school training in cultural ecology. Here was an autochthonous knowledge system that did not depend upon imported chemical fertilizers, costly fuel oil, and spare parts for pumps and tractors. It built instead upon the expertise and cumulative in situ landscape and agronomic knowledge of rice culture passed down through generations of women on land that remained under their control.

African Rice and Knowledge Systems in the Americas

When I first went to The Gambia, I wondered how an Asian crop had assumed such importance in West Africa. Why were West Africans growing rice and when had the Asian crop been introduced? I read that the Portuguese had introduced rice to the region during their maritime voyages. But in carrying out the library research for my dissertation, I was astonished to learn that rice had been in West Africa for several millennia. This was not *Oryza sativa*, the rice of Asian origin, but *O. glaberrima*, an indigenous species independently domesticated in the wetlands of Mali nearly four thousand years ago. From there, African rice diffused over a vast region, east to Lake Chad in the country by that name, west along Sahelian Rivers and inland swamps to the Atlantic coast and southward across mangrove-forested estuaries from Gambia to Côte d'Ivoire. The traditional rice-growing practices I had observed among Gambian women formed part of a broader legacy: a knowledge system seemingly rooted in antiquity that was responsible for numerous varieties adapted to drought, salinity, flooding and to specific agro-ecologies.

A few years into my position as assistant professor of Geography at UCLA, I came across historical research that attributed the origins of the rice plantation economy of South Carolina to enslaved West African rice growers.[11] I realized that my research on indigenous African rice agro-ecologies and practices could contribute to this scholarship.[12] I reviewed all available European commentaries on African rice systems over the first centuries of the transatlantic slave trade. These accounts provided considerable detail on the micro-environments cultivated as well as the prominence of women in the cereal's cultivation, marketing and milling. The historical overview complemented my earlier fieldwork by enabling an identification of the principal African rice-growing environments.

The result of this research many years later was *Black Rice: The African Origins of Rice Cultivation in the Americas* (Harvard University Press, 2001). The research started out by focusing on the South Carolina low country where historians had made a strong argument for African expertise in establishing an introduced crop that shaped the colony's rice plantation economy in the late seventeenth century. I then went through different archival and material evidence from other New World colonies to show how the appearance of rice cultivation was linked to the presence of enslaved growers. The research on Atlantic rice history eventually returned me to Brazil and fittingly, Maranhão, where the Portuguese in the mid-eighteenth century had launched a rice plantation economy, emulating that of the Carolina colony in North America. Rice had remained the dietary staple of their descendants, whom I had met in the region in the 1970s.

I began to think of ways to follow the cereal's introduction under different colonial experiences. Fortunately, US historians had so thoroughly combed the archives that I could identify the critical documents necessary for identifying the micro-environments planted and comparing techniques from field to kitchen with those described in West Africa. But this work also demanded considerable familiarity with the vast literature on slavery. I found a way forward by focusing attention on plantation food systems, crops grown for subsistence and the period when cultivated rice (an introduced crop to the Americas) first made its appearance. This research involved thinking about rice as a system of knowledge, the suite of practices that encompasses its planting, cultivation, harvesting and milling. I term the entirety of knowledge and practices 'rice culture' in order to bring into relief both agricultural methods as well as post-harvest processing, the way the grain is milled and cooked. The basic features of African rice culture then provided the methodology for a cross-cultural diachronic exploration of Atlantic rice history.

Cultural and political ecology again pointed the way for considering the flow of African knowledge systems in the context of forced migration, enslavement, subsistence choices and cultural identity. While my intellectual horizons at Berkeley originated with the insights of Carl Sauer and his respect for indigenous Amerindian knowledge systems, I had been struck in my own Latin American fieldwork by the seeming reticence of geographers to similarly engage the contributions of enslaved Africans in the Americas.[13] Working in rural areas of the Black Atlantic world made me want to address this historical lacuna. I wished to illuminate the role of New World Africans in establishing food plants of African origin in new biophysical and socio-economic environments. Their botanical legacy also involved recognition of plant genera of pan-tropical distribution known for their medicinal properties.[14]

While researching the onset of rice cultivation in northeastern South America (the Guianas and Brazil), I learned of the maroon oral history that tells how an enslaved African woman introduced rice by hiding grains of the cereal in her hair as she disembarked a slave ship. Here was a view of rice introduction to the Americas that stood in stark contrast to the bulk of scholarship on the Columbian Exchange. Coined by historian Alfred W. Crosby, the term Columbian Exchange refers to the Amerindian crops and European agency that transformed global food systems and environments between the sixteenth and nineteenth centuries. The Columbian Exchange literature says little about the African crops that were introduced to the Americas partly because not many people considered the possibility that African slaves might have their own forms of agricultural agency under severe, coercive circumstances.

The story of maroon rice provided independent testimony of the significance of African females for the diffusion rice culture across the Black Atlantic.

I extended this research in a second book which covered many more of the African food crops introduced to the Americas during the transatlantic slave trade.[15] I showed how Columbian Exchange scholarship could benefit from understanding its African components, how food grown in Africa for provisioning slave ships played a critical role in plantation societies when slaves carried and planted food gardens from whatever was left over or saved from the Atlantic crossings. Slaves labouring in plantations also cultivated sorghum, millet, yams and other foods from their continent in their subsistence gardens and created new food assemblages from what I like to think of as 'botanical gardens of the Atlantic world's dispossessed'.

Conclusion

In writing this chapter, I am struck by how tenuous much of the research seemed to me while I was in the field. Retrospection always tempts the memory to implant a grand, unified vision in terms of what one has seen and accomplished. In truth, I was unsure about where I was going on my academic journey into poverty, hunger and inequality. All I know was that the experience profoundly affected me, my view of the world, the paths I followed, and the scholars with whom I interacted and worked. It has been an enormous privilege to have the opportunity to live in different areas of the world and to know people from different cultures and perspectives. If I had my life to live over again, I would follow the same path because it led to discovery, self-awareness and compassion. The personal transformation that research engenders is not often evident to those who read the articles and books academics write. Theories provide the way scholars talk to each other, but the vitality of the experience that guides its formulation is often lost in communication.

In my lifetime, I have learned to appreciate the opportunity to travel outside my own culture, to ask large questions and to venture uncertainly in search of answers, even when conclusions were only partial and hypotheses seemed untenable. Along this journey, I had the good fortune to be among an exceptional cohort of graduate students at Berkeley and to have worked and collaborated with Michael Watts. The original work I did for CRED led to a dissertation and study of traditional rice ecologies that would inspire two books. Michael Watts gave me an opportunity to work in West Africa and the intellectual encouragement to pursue my research. Thank you, Michael, for the grand inspiration you gave to me, which turned a Latin Americanist into an Africanist, and an Africanist into a student of the Atlantic World. Your work in political ecology, development studies, on poverty and hunger, and on resource scrambles and petro-violence continues to inspire well beyond the academy.

Acknowledgments

The author would like to thank Jesse Ribot, Haripriya Rangan and Richard Rosomoff for their helpful comments on an earlier version of this chapter.

Notes

1   During the late 1970s the principal Latin Americanist professors in Berkeley Geography were Jim Parsons, Hilgard O'Reilly Sternberg and Barney Nietschmann; Clarence Glacken was also an inspiring presence through his concern with environmental history and intellectual modesty. Other notable campus Latin Americanist scholars included Woodrow Borah in History, Alain de Janvry in Agricultural Economics and Herbert Baker in Botany. Within Geography, graduate students working on environmental issues in Latin America at this time included Susanna Hecht, Nigel Smith, Karl Zimmerer and Bob Voeks.

2   FUNAI (Fundação Nacional do Indio or National Indian Foundation) is the Brazilian government agency charged with establishing and carrying out policies that concern the country's indigenous peoples, their lands and affairs.

3   During the transatlantic slave trade, Brazil received some four million slaves, about 40% of the total number of Africans forcibly migrated to the Americas. The plantation economy in Maranhão specialized in rice, cotton, indigo and coffee. Slavery was not formally abolished in Brazil until 1888, making it the last country in the Western hemisphere to do so.

4   With the return to democratic rule in 1985s, Brazil passed a law that enabled *quilombos* to qualify for land titles based on registering their land as collectively owned. Maranhão holds one of the highest *quilombo* concentrations in Brazil. Some 856 hamlets have petitioned for land titles, but only a few have in the decades since actually received them. See Sanzio Araújo dos Anjos (2009).

5   Geographers took the lead in developing political ecology as a research framework. Additionally influential were: Blaikie and Brookfield (1987) and Hecht and Cockburn (1989).

6   Sen (1981). Michael Watts' own substantial contribution had been on the political economy of hunger in northern Nigeria: Watts (1983).

7   Deere (1976); Agarwal (1982); Beneria and Sen (1982).

8   Gamble (1949, 1955); Haswell (1963); Weil (1972); Dey (1981).

9   Scholars whose work was pioneering in this regard included Guyer (1981); Whitehead (1981).

10   Carney and Watts (1990); Carney and Watts (1991).
11   Wood (1974); Littlefield (1981).
12   There are wild species of rice but only two were domesticated, one in Asia and the other in West Africa. These two species were introduced to tropical and subtropical America with the arrival of Europeans and Africans.
13   A notable exception was West (1957). James Parsons' research on the history of African grasses in tropical America, for instance, failed to consider this possibility. Parsons (1972).
14   Carney (2003).
15   Carney and Rosomoff (2009).

References

Agarwal, B. 1982. *Agricultural Modernisation and Third World Women*. Geneva, ILO.
Beneria, L. and Sen, G. 1982. Class and gender inequalities and women's role in economic development: Boserup revisited, *Feminist Studies*, vol. 8, no. 1, pp. 157–176.
Blaikie, P. and Brookfield, H. 1987. *Land Degradation and Society*. London, Routledge.
Carney, J. 2003. African traditional plant knowledge in the Circum-Caribbean region, *Journal of Ethnobiology*, vol. 23, no. 2, pp. 167–185.
Carney, J. and Rosomoff, R. 2009. *In the Shadow of Slavery: Africa's Botanical Legacy in the Atlantic World*. University of California, Berkeley.
Carney, J. and Watts, M. 1990. Manufacturing dissent: work, gender, and the politics of meaning in a peasant society, *Africa*, vol. 60, no. 2, pp. 207–241.
Carney, J. and Watts, M. 1991. Disciplining women? Rice, mechanization and the evolution of Mandinka gender relations in Senegambia, *Signs*, vol. 16 no. 4, pp. 651–681.
Deere, C. D. 1976. Rural women's subsistence production in the capitalist periphery, *The Review of Radical Political Economics*, vol. 8, no. 1, pp. 133–148.
Dey, J. 1981. Gambian women: unequal partners in rice development projects? *Journal of Development Studies*, vol. 17, no. 3, pp. 109–122.
Gamble, D. P. 1949. *Contributions to a Socio-Economic Survey of The Gambia*. London, Colonial Office.
Gamble, D. P. 1955. *Economic Conditions in Two Mandinka Villages*. London, Colonial Office.
Guyer, J. 1981. Household and community in African studies, *The African Studies Review*, vol. 24, no. 5, pp. 87–137.
Haswell, M. 1963. *The Changing Pattern of Economic Activity in a Gambian Village*. London, HMSO.
Hecht, S. and Cockburn, A. 1989 *Fate of the Forest*. London, Verso.
Littlefield, D. C. 1981. *Rice and Slaves*. Baton Rouge: Louisiana State University Press.

Parsons, J. J. 1972. Spread of African pasture grasses to the American tropics, *Journal of Range Management*, vol. 25, pp. 12–17.

Sanzio Araújo dos Anjos, R. 2009. *Quilombos—Geografia Africana Cartografia Étnica Territórios Tradicionais*. Brasilia, Mapas Editora & Consultoria.

Sen, A. 1981. *Poverty and Famine*, Oxford, Clarendon Press.

Watts, M. 1983. *Silent Violence: Food, Famine and Peasantry in Northern Nigeria*. Berkeley, University of California Press.

Weil, P. 1972. Wet rice, women, and adaptation in The Gambia, *Africana*, vol. 19, pp. 19–20.

West, R. C. 1957. *The Pacific Lowlands of Colombia*. Baton Rouge, Louisiana State University Press.

Whitehead, A. 1981. I'm hungry mum: the politics of domestic budgeting, in K. Young, C. Wolkowitz and R. McCullagh (eds), *Of Marriage and the Market*. London, CSE Books, pp. 88–111.

Wood, P. 1974. *Black Majority*. New York, Knopf.

2

Getting Back to our Roots

Integrating Critical Physical and Social Science in the Early Work of Michael Watts

Rebecca Lave

Introduction

Political ecology was built on a combination of critical social and physical science. Piers Blaikie (1985), Susanna Hecht (1985) and Michael Watts (1983a, 1983b, 1983c, 1983d, 1985) paired blistering critiques of environmental injustice with sober presentations of physical evidence, and built quantitative social science data into ferocious exposés of the politics of environmental science. Social science was clearly dominant; of political ecology's pioneers, only Hecht's work came close to balancing the physical and social. But looking back it is striking how integral biophysical data was in the formative work in our field, particularly given how peripheral natural science (and even quantitative social science) appears today.

This intellectual history raises some key questions. Why were physical science data employed so extensively in early political ecology? Why did that shift in the mid-1990s with the rise of post-structural political ecology? And finally, how can the integration of critical social and physical science at political ecology's roots inform scholarship today? To answer these questions, I focus on the exemplary work of Michael Watts.

Other Geographies: The Influences Of Michael Watts, First Edition. Edited by Sharad Chari, Susanne Freidberg, Vinay Gidwani, Jesse Ribot and Wendy Wolford.
© 2017 John Wiley & Sons Ltd. Published 2017 by John Wiley & Sons Ltd.

Early Political Ecology and the Integration of Critical Social and Physical Science

In the early 1980s, Watts published a set of publications that targeted dominant explanations of famine. Taking on 'tragedy of the commons' (Watts 1985), Malthusian (Watts 1983a, 1983b, 1983d, 1985), and 'stupid peasant' arguments (Watts 1983a, 1983b, 1983d, 1985), he argued, in Peet and Watts' (1996, 4–5) subsequent description, that

> environmental problems in the Third World ... are less a problem of poor management, overpopulation, or ignorance, as of social action and political economic constraints ... [Analysis should thus concentrate on] market integration, commercialization, and the dislocation of customary forms of resource management...

Watts opened his arguments by describing how famine in Northern Nigeria had achieved, 'a sort of textbook notoriety and is usually invoked as an archetypical example of poor land use, desertification, local or global climate change, neo-Malthusian population pressure, or life boat ethics in practice' (Watts 1983b, 24). Like Blaikie and Hecht, one of Watts' primary goals was to expose the politics underlying these supposedly neutral explanations. In an eloquent early version of the now classic argument that there is no such thing as a natural disaster, Watts (1983b, 26) argued that

> natural hazards are not really natural, for though drought may be a catalyst or trigger mechanism in the sequence of events which leads to famine conditions, the crisis itself is more a reflection of the ability of the socio-economic system to cope with the unusual harshness of ecological conditions and their effects. To neglect this fact is to ... [miss] a major political point.

The actual cause of famine, Watts argued, was political economic inequality, not bad management of resources, overbreeding or stupidity. Instead of 'simple technical or demographic failures or ... the inevitable consequence of a predatory climate' (Watts 1983a, xx), twentieth-century famines in Northern Nigeria should be explained by 'the integration of pre-capitalist forms of production into a global capitalist system, largely under the aegis of the colonial state' (Watts 1983b, 26). Thus the stakes in debunking dominant explanatory frameworks were, very clearly, as political as they were intellectual.

Watts' critique of conventional explanations of famine in Sudano-Sahelian Africa was deeply historically grounded. For example, he demonstrated the fallacy of Malthusian explanations via quantitative historical data: the Sokoto Caliphate maintained dense populations (up to 300/mi^2) in the area throughout most of the nineteenth century under similar climatic conditions with recurrent food availability crises, but without great famines (Watts 1983a, 61). Overpopulation was clearly not the cause of twentieth century famines in Northern Nigeria.

Neither was peasant ignorance. As Watts demonstrated (drawing on historical climate data) drought, and thus hunger, had been a recurrent and powerfully formative part of life in Northern Nigeria for centuries (Watts 1983a, 1983b, 1983c, 1983d). Given that the semi-arid climate made 'average' annual rainfall a statistical fiction rather than a lived reality, generation after generation of Hausa farmers had been confronted with substantial variation in rainfall quantity, geography and timing, all of which imperilled crops and thus required sophisticated agronomic responses. As Watts (1983a, 109–112) described it:

> There are three obvious attributes of arid and semiarid ecosystems; first, precipitation is so low that water is the dominant controlling factor for biological processes; second, precipitation is highly variable through the year and occurs in discrete, discontinuous packages or 'pulses'; and third, spatiotemporal variation in precipitation has a large random (unpredictable) component. From an anthropocentric perspective the water-controlled nature of arid ecosystems, principally due to the tight coupling of energy inflow with water inflow, translates into agropastoral systems that are highly vulnerable to drought and rainfall uncertainty ... Precipitation is almost never 'normal' over the short or long term; it is almost always considerably above or below any mean statistic.

Farmers were thus 'acutely concerned with the concrete, empirical variability in annual rainfall, and agronomic practice varies in tandem with the precise pattern of precipitation' (Watts 1983d, 56). Given farmers' deep experience with and knowledge of agronomy in response to climatic variation, Watts argued, famine can only occur if farming families are 'constrained by their poverty and more generally by the structure of village level political economy. These households are actually *prevented* from responding adequately ...' (Watts 1983d, 66, emphasis in original) by political economic relations. British colonial rule and its accompanying emphases on monetization and commodity production undermined the informal social structures of the moral economy that had substantially reduced the risk of famine in times of drought for centuries, but did not replace them with an effective safety net. Thus explaining the great famines of the twentieth century based on peasant ignorance was just as absurd as blaming it on overpopulation.[1]

One of the striking features of Watts' approach in these early texts was its incorporation of quantitative social and physical science data into radical qualitative social analysis. This combination was unusual at the time, and is quite different from most research that sails under the political ecology flag today (but see Turner 2015). What drove it?

One factor was clearly the theoretical framing of early political ecology, which was grounded primarily in Marxist political economy. This theoretical choice carried important ontological, epistemological and methodological consequences.[2] While Marx's relatively sparse writings on nature/society relations have been interpreted, in Talmudic fashion, to support a wide range of arguments (contrast, for example, Engel Di-Mauro 2014 with Robertson and Wainwright 2013), Watts seems to have understood them as an argument for the co-constitution of nature and society, as demonstrated in what follows. Unlike the constructivist epistemologies that later came to dominate political ecology, Marx insisted that it was possible to diagnose the true character of social relations. While that truth was hidden by power relations characteristic of the dominant mode of production, a careful analyst (i.e. Marx) could peel back the veil to reveal objective truth. This in turn meant that Marx was not hostile to the realist basis of natural science truth claims. Although he himself did not conduct natural science research, he was keenly engaged with central ideas in natural science, from evolution to soil chemistry, and incorporated them as key arguments in his writing.

Watts' approach in his early work on famine is shaped by similar ontological, epistemological and methodological commitments. The world Watts describes is one in which the natural and the human are not neatly separated categories but co-constituted entities. So for example, in contrast to then dominant explanations of famine, he argued that '[f]amine is simultaneously a biological and social experience. Its etiology may be as profoundly economic as environmental and its effects as much political as physiological' (Watts 1983a, 13). More generally, Watts argued that '[w]e clearly live not only in an environment constituted by natural processes but also in one of our making, socially constituted by human practice and subject to ongoing change and historical transformation' (Watts 1983a, 14). Put differently (Watts 1983a, 87–88):

> [N]ature separate from society has no meaning. This is not simply to suggest that nature is mediated through, and related to, social activity but rather that, in both historical and practical senses, nature resides at the locus of all human practice ... There is ... an irreducible unity between society and nature that is differentiated from within.

Beyond these ontological roots in Marx, Watts incorporated natural and social science data into his work (as his foregoing statements on climate indicate) as important claims to truth, demonstrating an approach to knowledge and methodological breadth reminiscent of Marx. Thus the unusual combination of critical social and physical science characteristic of Watts' early work (and of early political ecology more generally) should be ascribed in part to the effects of political ecology's initial grounding in Marxist political economy.

A second factor behind the use of physical science data by Watts and other early political ecologists seems to have been the political project driving their research. The 'stupid peasant', Malthusian and 'tragedy of the commons' explanations that dominated international environmental policy at the time were 'supported' with references to colonial-era physical science. To debunk arguments that nomads' irrational and ignorant herding practices were responsible for desertification, for example, it was not sufficient to demonstrate that herder societies were deeply knowledgeable about interrelationships between ecological conditions and their livestock practices. Watts had to be able to discredit assertions that desertification was actually occurring, and this required physical science data on climate and ecology. Blaikie (1985) and Hecht (1985) made similar intellectual moves to debunk prevailing claims about soil degradation. For early political ecologists, successfully taking on dominant policy regimes required both directly challenging the physical science 'data' on which they were based in order to expose them as colonialist, racist, classist propaganda, and replacing with data they considered more accurate. As in much Marxian scholarship, early political ecologists believed that while a good deal of what passed as science was deeply ideological, it was possible to put forward true scientific claims on the basis of careful analysis.

The Post-Structural Turn

In the 1990s, political ecology's dominant theoretical framework began to shift in what is typically referred to as the 'post-structural turn', and with it political ecologists' empirical focus transferred to the realms of social movements and discourse. The initial proclamation of this shift came in the Peet and Watts' germinal 1996 collection, *Liberation Ecologies*, which extolled political ecology's new intellectual influences, 'drawn from poststructuralism, gender theory, critical theories of science, [and] environmental history' (Peet and Watts 1996, 9). These new engagements with feminism, Foucault, social movements theory, and science and technology studies (STS) broadened the theoretical and methodological palette of political ecology research and expanded its empirical focus. For the purposes of this chapter, I focus primarily on the epistemological implications.

The post-structural turn as heralded by Peet and Watts involved an explicit reconsideration of the materialist epistemology characteristic of initial formulations of political ecology. The very first sentence of their discussion of poststructuralism in the Introduction to *Liberation Ecologies* signals this clearly: 'Poststructural theory's fascination with discourse originates in its rejection of modern conceptions of truth' (Peet and Watts 1996, 13). And later (Peet and Watts 1996, 15–16):

> We find these positions attractive in that ... poststructural theory links with ... the relentless critique of everything, even notions usually considered to be emancipatory ... By criticizing the modern belief in rational humans speaking objective science, poststructural theory opens a space in which a wide range of beliefs, logics, and discourses can be newly valorized.

Accurately describing what has turned out to be the intellectual trajectory of political ecology for the subsequent two decades, Peet and Watts state that 'post-structural concerns with knowledge-power, institutions and regimes of truth, and cultural difference have proven compelling in the rethinking of ... political ecology' (Peet and Watts 1996, 2), resulting in a fundamental shift in empirical focus from the *materiality* of nature to its *representation* in dominant discourses. While there are certainly exceptions to this, political ecology's engagement with natural science today is characterized best not by critical ecology (or pedology or hydrology), but instead by 'critical approaches to ecological science itself' (Peet and Watts 1996, 13).

While this call to critique science and materialist epistemologies is what is best remembered from *Liberation Ecologies*, it is worth noting that the relationship to physical science laid out in Peet and Watts' Introduction is more complex, and not particularly constructivist. I would point to two moments in particular that complicate the common characterization of the positions staked out in the Introduction. First, while the authors do indeed call for 'interrogation of the term "ecology" in political ecology,' this is not because earlier work relied on physical science *per se*, but because it relied on *incorrect* physical science, 'a rather outdated view of ecology rooted in stability, resilience, and systems theory' (Peet and Watts 1996, 12). Because of (Peet and Watts 1996, 12)

> [t]he shift from 1960s systems models to the ecology of chaos, that is to say chaotic fluctuations, disequilibria, and instability, ... many previous studies of range management or soil degradation resting on simple notions of stability, harmony, and resilience may need to be rethought.

A second key moment that belies the typical, strongly constructivist reading comes in the last section of the Introduction, when Peet and

Watts state that '[o]ur own [epistemological] position... tends towards a critical modernism in which rationality is contended rather than abandoned' (Peet and Watts 1996, 37–38). This is a rather different epistemological bent than is typically attributed to *Liberation Ecologies*' highly influential Introduction. Epistemological ambivalence towards post-structuralism is visible in the body of *Liberation Ecologies* as well. While the topical focus on race, gender, social movements and discourse provides a unifying structure for the volume, it seems to have masked the collection's epistemological breadth: there is considerable difference between Judith Carney's and Karl Zimmerer's distinctly realist epistemologies and the constructivism of Arturo Escobar and Jake Kosek.

This suggests that it is worth separating the diverse intellectual approaches casually lumped together as 'the post-structural turn'. In my view, the important deepening of political ecology's engagement with race, gender, social movements, and discourse was neither dependent on the embrace of constructivist epistemologies nor on the methodological shift away from natural science and quantitative data more broadly, as authors such as Fairhead and Leach (1996) (and indeed early political ecology itself) conclusively demonstrate. Paul Robbins (2015) persuasively argues that political ecologists have often wanted to 'have it both ways', critiquing colonialist ideological science while conducting physical science research of their own that they take as producing truth. In his pithy phrasing, political ecology's common project is to 'draw into doubt scientific accounts of environmental conditions or change while proliferating them' (Robbins 2015, 92). This is a quintessential expression of the materialist epistemology characteristic of early political ecology.

And yet despite the fact that many central aspects of the post-structural turn were not inherently dependent on strongly constructivist epistemologies – and thus, on the methodological abandonment of physical science data – political ecology had lost most of its active engagement with physical science by the end of the 1990s. There were a few exceptions, most notably the nature/society group at University of Wisconsin-Madison, where Matt Turner and Karl Zimmerer (and more recently Paul Robbins) held open a space for the ecology in political ecology (Turner 2015). But political ecology research is no longer characterized by its fusion of critical social and physical science.

New Engagements: Critical Physical Geography

Looking back at Watts' early work from a vantage point three decades later, what do we see? What might we draw from initial formulations of political ecology to inspire current research? I would point in particular to two qualities: intellectual breadth and methodological pluralism.

The intellectual breadth of political ecology research in the early- and mid-1980s is striking, encompassing social and physical science. This breadth gave Watts, Blaikie, Hecht and their compatriots the epistemological remit to chase environmental issues across their full human and physical range rather than having to rein in pursuit at the academic border.

This intellectual breadth was enabled by a fluid combination of qualitative and quantitative data, which allowed early work in political ecology to build impressively robust claims. In *Silent Violence*, for example, Watts moved smoothly between *qualitative social science* data drawn from ethnographic fieldwork and archival materials; *quantitative social science* data drawn from surveys and archival sources; and *quantitative and qualitative physical science* data drawn from oral histories, archival data and research into changing climate patterns in Sub-Saharan Africa. This provided a formidable and sturdy base for his arguments, giving early political ecology an explanatory power and level of credibility that more recent work lacks in the eyes of physical scientists and more quantitatively focused policy makers. Again, I want to note that there are scholars who continue to work in this initial integrative vein; but current research in political ecology only rarely strays far from qualitative social science. This allows it to explore particular social and discursive formations in great depth and richness. But there is another sort of treasure to be found in breadth, and it is here that the field of critical physical geography finds its jumping-off point.

Critical physical geography (CPG) is a rapidly growing body of work that draws together critical human and physical geography to study eco-social systems. CPG research combines 'critical attention to relations of social power with deep knowledge of a particular field of biophysical science … in the service of social and environmental transformation' (Lave et al. 2014, 2). Recent work includes critical examination of the political economy of lead distribution in urban soils (McClintock 2015), of the social construction of flood risk (Lane 2014), of the profoundly capitalist values embedded in soil science (Engel-Di Mauro 2014) and of the physical ramifications of ecosystem service markets for streams (Doyle et al. 2015). These articles, like other CPG scholarship, present and analyse primary physical and social data rather than relying on secondary sources for the former, as most early political ecology did.

The distinguishing features of this research are clearly not topical; as with political ecology, CPG researchers tackle a broad range of issues. Instead, what makes CPG distinct from other approaches that integrate physical and human geography is its careful attention to both the co-constitution of landscapes and power relations, and to 'the politics of environmental science and the role of biophysical inquiry in promoting social and environmental justice' (Lave 2015, 571). The agenda here is thus intellectual – integrating physical and critical social science; and political – examining the politics of knowledge production, but also promoting the emancipatory possibilities of that knowledge. For me and for many other CPG researchers, both of these core commitments have their roots in early political ecology (although it is important to note that CPG pays far more attention to biophysical processes and the material condition of landscapes than any of the early political ecologists, with the possible exception of Hecht).

What, exactly, does CPG enable us to do that other approaches do not? I would point to two key aspects. First, while the Anthropocene concept is problematic for any number of reasons (Haraway 2015; Moore 2015), it makes the necessary ontological point that the landscapes we study are now eco-social systems, powerfully shaped by both physical and human influences. Taking this seriously requires a fundamental epistemological and methodological shift in the practices of environmental researchers, since limiting ourselves to either physical or social analysis alone is clearly inadequate. CPG is one useful transdisciplinary approach, as it enables us to ask questions that more directly reflect current environmental conditions, and thus to study the 'crappy landscapes' we produce (Urban, in review) and human attempts to fix them via ecosystem service markets, designed ecosystems, restoration and rewilding.

The second point is integrally related to the first: a CPG approach gives our work greater explanatory power. In the same way that Blaikie, Hecht and Watts depended on quantitative social and natural science data to persuasively critique Malthusian and stupid peasant arguments, critical nature/society scholars today need quantitative physical and social evidence to successfully critique the dominant policy frameworks of the early twenty-first century, such as natural capital and market-based approaches more broadly. Triangulating between qualitative and quantitative natural and social science data allows CPG researchers to develop more robust findings, as well as to speak more persuasively to those who privilege quantitative – particularly quantitative natural science – explanations, such as policy makers.

Conclusions

We tend to give relatively short shrift these days to the initial texts that crystalized political ecology. It is a rare syllabus that devotes more than a week to work from the early and mid-eighties before rushing onwards to the post-structural turn with a nearly audible sigh of relief. The broader theoretical and methodological palette of political ecology research that arrived with the post-structural turn is clearly a very good thing for the field, expanding our intellectual and political reach to better address race, gender, social movements, and the role of discourse in environmental politics, all of which (with the possible exception of social movements) were included in early political ecology but took nowhere near as central an analytical position as they do today.

I have argued here, however, that the early work in our field had important strengths that we have abandoned, perhaps unnecessarily. Watts' writing from the early- to mid-eighties, and Blaikie's and Hecht's as well, demonstrated a striking methodological fusion of critical quantitative and qualitative social and physical science; an epistemological approach that critiqued politically driven knowledge claims while still insisting that researchers could peel back the ideological layers to find true explanations; and an ontology that rejected nature/culture dualisms. Contemporary political ecologists share the ontological commitments of early political ecology, but have departed quite strongly from the early scholarship's methodological and epistemological positions. Today, political ecologists primarily employ qualitative social science methods and constructivist epistemologies in which debates among knowledge claims cannot be settled on grounds of truth. The emerging field of critical physical geography, by contrast, builds upon both, drawing on the methodological, epistemological and ontological strengths of early political ecology, while also incorporating the expanded topical foci that give contemporary political ecology such appeal.

Should contemporary political ecologists re-engage with their field's early roots? I suspect every political ecologist will have to answer that question for themselves, based on their own intellectual and political commitments. Certainly, if we were to re-engage with political ecology's roots then Watts' early scholarship provides an exemplary entry point. But even if re-engagement is not on the agenda, it would be worthwhile to revisit *Liberation Ecologies* (Peet and Watts 1996) for a more nuanced reading of what that collection actually does (or does not do) for the post-structural turn, and to reconsider the epistemological and methodological commitments that have flowed from what I believe to be a somewhat oversimplified read of that volume.

Notes

1   There are obvious parallels here with Amartya Sen's powerful and highly influential work on famine, which was published just before the set of Watts' texts described earlier (Sen 1980, 1981). Yet while there are a handful of

references to Sen's work in *Silent Violence* (Watts 1983a), there is no extended engagement. It does not seem that the two were in conversation. As Watts said in a recent interview, 'It was one of those occasions when there were lots of people thinking about broadly very similar issues [in tandem]' (personal communication 20 July 2015). It is not clear whether this was sheer coincidence, or whether there were earlier texts that led them down shared paths; it would be very interesting to investigate this further.

2   The paragraph that follows is far too brief to address the complexities and nuances of Marx's work. The focus of this volume is Watts, not Marx, though, so I hope the Marxologists among my colleagues will forgive my cursory treatment.

References

Blaikie, P. 1985. *The political economy of soil erosion in developing countries.* New York, John Wiley & Sons.

Carney, J. 1996. Converting the wetlands, engendering the environment: the intersection of gender with agrarian change in Gambia, in R. Peet and M. Watts (eds), *Liberation Ecologies: Environment, development, social movements.* London, Routledge.

Doyle, M., Singh, J., Lave, R., and Robertson, M. 2015. The morphology of streams restored for market and non-market purposes: insights from a mixed natural-social science approach, *Water Resources Research*, vol. 51, no. 7, pp. 5603–5622.

Engel Di-Mauro, S. 2014. *Ecology, Soils, and the Left.* New York, Palgrave Macmillan.

Fairhead, J. and Leach, M. 1996. Rethinking the forest-savanna mosaic: colonial science & its relics in West Africa, in M. Leach and R. Mearns (eds), *The Lie of the Land: Challenging received wisdom on the African environment.* Oxford, James Currey.

Haraway, D. 2015. Anthropocene, Capitalocence, Chthulucene: making kin, *Environmental Humanities*, vol. 6, pp. 159–165.

Hecht, S. 1985. Environment, development and politics: capital accumulation and the livestock sector in Eastern Amazonia, *World Development*, vol. 13, no. 6, pp. 663–684.

Lane, S. 2014. Acting, predicting and intervening in a socio-hydrological world, *Hydrology and Earth Systems Sciences*, vol. 18, no. 3, pp. 927–952.

Lave, R. 2015. Introduction: critical physical geography, *Progress in Physical Geography*, vol. 39, no. 5, pp. 571–575.

Lave, R., Wilson, M. W. and Barron, E. 2014. Intervention: critical physical geography, *The Canadian Geographer*, vol. 58, no. 1, pp. 1–10.

McClintock, N. 2015. A critical physical geography of urban soil contamination, *Geoforum*, vol. 65, pp. 69–85.

Moore, J. 2015. *The Capitalocence part I: on the nature & origins of our ecological crisis.* http://www.jasonwmoore.com/uploads/The_Capitalocene_Part_I_June_2014.pdf (accessed 1 December 2015).

Peet, R. and Watts, M. (eds). 1996. *Liberation Ecologies: Environment, development, social movements*. London, Routledge.

Robbins, P. 2015. The trickster science, in T. Perreault, G. Bridge, and J. McCarthy (eds), *The Routledge Handbook of Political Ecology*. New York, Routledge.

Robertson, M. and Wainwright, J. 2013. The value of nature to the state, *Annals of the American Association of Geographers*, vol. 103, no. 4, pp. 890–905.

Sen, A. 1980. Famines, *World Development*, vol. 8, no. 9, pp. 613–621.

Sen, A. 1981. *Poverty and Famines: An essay on entitlement and deprivation*. Oxford, Clarendon.

Turner, M. 2015. Political ecology II: engagements with ecology, *Progress in Human Geography*, vol. 40, no. 3, pp. 413–421.

Urban, M. In review. In defense of crappy landscapes, in R. Lave, C. Biermann, and S. Lane (eds), *The Palgrave Handbook of Critical Physical Geography*. London, Palgrave MacMillan.

Watts, M. 1983a. *Silent Violence: Food, famine, and peasantry in Northern Nigeria*. Berkeley, University of California Press.

Watts, M. 1983b. Hazards and crises: a political economy of drought and famine in Northern Nigeria, *Antipode*, vol. 15, no. 1, pp. 24–34.

Watts, M. 1983c. On the poverty of theory: natural hazards research in context, in K. Hewitt (ed.), *Interpretations of calamity from the viewpoint of human ecology*. Boston, MA, Allen & Unwin.

Watts, M. 1983d. The political economy of climatic hazards: a village perspective on drought and peasant economy in a semi-arid region of West Africa, *Cahiers d'Etudes Africaines*, vol. 23, no. 89/90, pp. 37–72.

Watts, M. 1985. Social theory and environmental degradation: the case of Sudano-Sahelian West Africa, in Y. Gradus (ed.), *Desert Development: Man and technology in sparselands*. Dordrecht, Netherlands, Reidel.

3

Binary Narratives of Capitalism and Climate Change

Dangers and Possibilities

Lucy Jarosz

I am interested in the melding of human and nature.
Chiho Aoshima, Artist

Environmental crisis narratives about the relationship between humans and nature tell of human-induced environmental degradation, destruction and change and of the impacts and responses to these changes. Binaries, such as human/nature and local/global, structure many of these influential crisis narratives. Binaries have been critiqued by political ecologists and eco-feminists, because they cannot adequately explain or help us understand the dynamics of environmental destruction, degradation and change in specific places, regions and across history and scale. Binaries can also oppress and exploit humans and nature through hierarchies, discrimination and other-ing. But binaries are foundational in environmental crisis narratives, because they can energize, motivate and mobilize the political will to action. For example, Rachel Carson's *Silent Spring*, with its attendant binaries of life/ death and good/evil, is widely recognized as ushering in the American environmental movement of the 1970s, playing a role in the creation of the Environmental Protection Agency, and in the subsequent ban of DDT spraying in the United States. Rather than reject binaries entirely, I argue that strategically deploying them can work to shift dominant neoliberal and cultural narratives about people's relationships to nature.

Other Geographies: The Influences Of Michael Watts, First Edition. Edited by Sharad Chari, Susanne Freidberg, Vinay Gidwani, Jesse Ribot and Wendy Wolford.
© 2017 John Wiley & Sons Ltd. Published 2017 by John Wiley & Sons Ltd.

Political ecology's dialectical approach delineates the specific complexities and paradoxes evident in a deep and detailed analysis of the political economy/ecology of environmental change (see Harvey 1996 on dialectics). This approach eschews binaries, because they cannot accurately explain the complex and messy realities of struggle and transformation bound up in environmental politics. In his review of Naomi Klein's environmental crisis narrative, *This Changes Everything: Capital vs. Climate*, Michael Watts (2015, 111) concludes by asking whether the politics of climate change 'really turns on the sorts of binaries Klein lays out'. He notes that research about the politics of capital and climate change reveal paradox, ambiguity, relationality and coincidence. He rightly points to the dangers of employing binaries to understand and explain the complex material and representational geographies of climate change in places such as the Niger Delta. Environmental crisis narratives are polemical calls to action and resistance. They tell of a relationship between people and nature defined by human domination and the looming destruction of both people and the planet.

Binaries can play a vital role in cultural narratives if, by conveying urgency, they mobilize political will and lead to actions to protect the environment and recognize the most vulnerable social groups. Yet the forms of social activism that binaries unleash can also be oppressive if, like certain neo-Malthusian crisis narratives have done, they stoke the flames of racism, xenophobia and sexism. These divergent possibilities make binaries fascinating discursive tools and signs. They can narrate, politicize and obscure place-based events, processes and situations, and their deployment can also yield arguments for interconnection, caring and holism. While I agree with Watts that binaries do not accurately depict on-the-ground environmental politics, I think that their primary purpose lies elsewhere. Binaries do important work in raising the public awareness of key environmental and public health issues needed to motivate and sustain activism. They can be dangerous, but they can also help us envision caring, post-capitalist geographies (Gibson-Graham 2006; Lawson 2009). The binaries employed in environmental crisis narratives serve to reveal and politicize harmful forms of environmental change, while political ecology attends to the materiality and complexity of that change in particular places. Both theoretically and empirically, political ecology helps to decentre and trouble binaries. But what if we begin to trouble and decentre the boundaries between political ecology and environmental crisis narratives? What would be the implications for theory, knowledge production and activism?

In short, we need both political ecology and environmental crisis narratives in order to analyse, interpret, understand and actively promote justice and equity. In this chapter I explore how binaries are deployed in

Naomi Klein's polemic on climate change, *This Changes Everything*, and how they are challenged by Michael Watts' analysis of African famine and by feminist political ecology. I also want to show how binaries are increasingly refused by theoretical approaches and cultural narratives that envision non-binary relationships between society and nature. We see this in research on embodiment and queer ecologies, for example, as well as in the artwork of Japan's Chiho Aoshima and the oral narratives of the San people of South Africa. This melding of human/and nature is also relevant to political ecology, which revealed how impoverished and vulnerable people and places were made. Soil erosion was both biophysical process and rooted in a political economy (Blaikie 1985). Hunger and famine are silent violences that are produced by climate and by economic and political processes (Watts 1983).

Silent Violence: The Persistence of Hunger and Famine

In the introduction to the second edition of *Silent Violence*, Watts writes that 'the deadweight of global warming is about to fall' (2013, xliii). For him it is what Gramsci called an interregnum: a time and space when the old order is dying and the new is yet to be. For others, such as journalist Christian Parenti, climate change is already fuelling violence and famine across regions constituting what he describes as 'the tropic of chaos' (Parenti 2011). Parenti's imagined geography is reminiscent of other binaries, such as those drawn in Robert Kaplan's (1994) 'The Coming Anarchy' and Homer-Dixon's (1999) depiction of a world destabilized by war, hunger and outmigration in the global South. These crisis narratives deploy binaries such as order/chaos, North/South and them/us harking back to the white civilization/dark continent binaries resonating with Joseph Conrad's *Heart of Darkness* (Jarosz 1992). While Watts also envisions a dark and difficult future, his explanation roots it in the historical geography of colonization and agrarian capitalism, rather than in the population growth, scarcity and conflict which are themselves outcomes of these political-economic processes.

One of the significant contributions of *Silent Violence* is that it links hunger to struggles over access to and control of land, water, technologies and labour, while also mapping these struggles onto the broader historical geography of colonialism and capitalist development. It is now widely accepted that famines and hunger have political and economic roots and that both environment and society actively constitute one another through time and place. Writing about food crises among Hausa peasant farmers in northern Nigeria, Watts made these arguments over thirty years ago. *Silent Violence* also countered neo-Malthusian views that attribute hunger and famine to some combination of 'overpopulation' and natural hazards, such as drought. He (Watts 1983, 17) writes: 'Drought does not necessarily cause famine, as much of the discourse surrounding the Sahelian famine of the 1970s implied ... much of what passes as natural hazards are not really natural at all ...' A central question for political ecology at that time was how environmental degradation and change was socially, politically, culturally and economically produced. As one of Watts' graduate students in the late 1980s, I was drawn to *Silent Violence* because, among other things, it pushed back against the them/us, North/South, developed/underdeveloped binaries I was reading about in popular environmental literature, which saw poverty and environmental degradation as a product of brown and black people's unproductive, traditional modes of farming and food provisioning.

Naomi Klein assumes that the realities of climate change are also rooted in the political economy of natural resource extraction and globalization. She argues that capitalism's market-based solutions cannot respond effectively to what she sees as 'a clear and present danger to civilization', due to capitalism's role in driving the exploitation of both people and natural resources. According to Klein (2014, 23), '... the triumph of market logic, with its ethos of domination and fierce competition, is paralyzing almost all serious efforts to respond to climate change'. But in contrast to Watts, she (Klein 2014, 22) represents climate change as a battle between capitalism and the planet, thereby assuming a society/nature binary common to environmental crisis narratives such as Rachel Carson's *Silent Spring* and Barry Commoner's *The Closing Circle*.

Binaries in Cultural Narratives of People and Nature

Binaries such as society/environment, capital/climate and local/global, among others, mark many key texts of environmental crisis and change since the 1970s. These binaries often structure cultural narratives about people and nature in apocalyptic terms of resource scarcity, environmental destruction and death (Ehrlich 1968; Carson 1962; Commoner 1971; Cribb 2010). While they simplify and obscure complex processes and events, their creative expression may evoke powerful emotions and understandings. For example, using the binary of life/death, Rachel Carson writes of how chemical poisonings bring a silent spring to the American rural heartland where previously birds sang and 'foxes barked in the hills and deer silently crossed the fields, half hidden in the mists ... (Carson 1962, 1). Chemicals bring death to people and to nature, and Carson calls for abandoning the ideology of the human domination of nature through research and technology and replacing this binary with understandings and practices that work in harmony with nature rather than dominating and destroying it and ultimately ourselves.

For over three centuries, the Western human/nature binary in particular has conceived of a hierarchy in which humans dominate, degrade and destroy nature and ultimately themselves. Some see this hierarchal binary underpinning the human contribution to climate change, and call for it to be challenged (Hamilton 2011). Feminist political ecology has long done so: in *Death of Nature* (1980), Carolyn Merchant argued that the society/nature binary was also expressed in the man/woman binary and, that the destruction of nature was deeply linked to the exploitation and the oppression of woman. More recently, environmental philosopher Val Plumwood, drawing on critiques from post-structuralism, postmodernism and eco-feminism (Plumwood 1993, 31) argues that we need to rethink our basic assumptions about the human ability, necessity and mandate to dominate nature (2009, 53, 65).

Human/nature binaries portray a divide that is both essential and unchanging in which humans dominate, tame and subdue nature. They tap centuries-old Western cultural constructs that see the human as part of a radically separate order of reason, mind or consciousness, set apart from a lower order of nature. They neglect evidence of interconnectedness and embeddedness. By contrast, environmental historians (Nash 2006) and political ecologists (Guthman 2011) conceptualize the boundaries between human and nature as highly porous. Public health problems become inseparable from environmental ones, as in the case of chemicals and toxins in food supplies and in soil and water. Political ecologists embed people and nature in relations of dependency and mutuality, not hierarchy, questioning even the purity of scientific categories (Biermann and Mansfield 2014). For Plumwood (1993), the human/nature binary shapes and distorts identities on both side of the dualism. She (1993) argues that binaries do not encourage or sometimes even allow for reconciliation.

Binaries in *This Changes Everything*

Klein's central binary is the subtitle of the book: Capital versus Climate. She argues that neoliberal capitalism, as exemplified by international trade treaties and WTO rules, is incompatible with the transition away from fossil fuel development toward more low-carbon, sustainable and just economies. She starts with the contemporary moment when deregulation, lower corporate taxation and public spending cuts have contributed to market instability, rising levels of inequality, poverty and hunger, and the erosion of public infrastructure and services (Klein 2014, 19). As the source of these problems, Klein (2014, xx) argues, capitalism cannot offer solutions to climate change. She calls instead for a transition away from a careless economy towards one that would foster more caring social relationships among humans and with the nonhuman world. Thus the binary of corporate capital versus society and nature outlines the problem while the careless versus caring economy describes one solution. A third binary pits grassroots social movements against state, corporations, and elites, as illustrated by resistance to mining and fossil fuel-based development such as the Keystone XL pipeline system. Together these three binaries express what Klein (2014, 22) identifies as 'the battle between capitalism and the planet.'

Klein (2014, 294) defines a space of struggle as *Blockadia*, which is a 'roving transnational conflict zone' that appears in specific locations where fossil fuels are extracted, refined or transported. Resistance to extraction is both global and grassroots and Klein briefly notes grassroots resistance movements from Greece to Canada to the Amazon and the Niger Delta. These resistance movements are deeply place-based, and Klein calls for alliances among indigenous peoples and nations, small-scale farmers, pastoralists, urban and small town community and social activists, fishers and gardeners, and environmentalists reminiscent of alliances formed during the WTO protests in Seattle in 1999 and elsewhere. These alliances exist around a plurality of issues including poverty and hunger, agriculture and food, human rights, social and environmental justice, and freedom and liberation, among others (Klein, 2002). They are part of the anti-globalization movement crystallized in the WTO protests and in the Occupy Movement. NGOs working on behalf of climate justice, small-scale farming and sustainable agriculture and social justice are a part of this movement as well.

Klein also invokes the now familiar global versus local food binary (see also Norberg-Hodge 2002). Here the problem is industrial agriculture, and the envisioned solution is smaller scale farms, practicing organic or agro-ecological techniques within democratically controlled and equitable food systems (Klein (2014, 136). Informing this vision is Miguel Altieri's (2011) research on agroecology, which shows how knowledge created and shared among *campesinos* (farmers or peasants) has both enriched on-farm ecological diversity and helped them practise healthy and climate friendly food provisioning.

Binaries can mark the sides of struggle, whether around issues of food justice or self-determination. Ultimately, however, they indicate the need to rethink the human/nature binary, with its connotations of domination and separation, as a more dialectical relationship of interconnection, dynamic change and interdependence. As Watts indicated, binaries do not have explanatory or analytic power, but they are narrative devices that are frequently used to define crises and delineate political positions and spaces. Rather than rejecting binaries entirely, I suggest asking how they might allow for reconciliation, alliances, and new relationships. How might binaries be strategically deployed in the construction of counternarratives (Goodall 2010)? The strategic use of binaries can offer possibilities for knowledge construction around climate change, while at the same time we must, as Watts noted, be sceptical of how they frame and explain realities.

References

Altieri, MA and Toldeo, VM 2011 The agroecological revolution in Latin America: rescuing nature, ensuring food sovereignty and empowering peasants, *Journal of Peasant Studies*, vol. 38, no. 3, pp. 587–612.

Biermann, C. and Mansfield, B. 2014. Biodiversity, purity, and death: conservation biology as biopolitics, *Environment and Planning D*, vol. 32, no. 2, pp. 257–273.

Blaikie, P. 1985. *Political Economy of Soil Erosion*. London and New York, Longman.

Carson, R. 1962. *Silent Spring*. New York, Houghton Mifflin.

Commoner, B. 1971. *The Closing Circle*. New York, Alfred A. Knopf.

Cribb, J. 2010. *The Coming Famine*. Berkeley, University of California Press.

Ehrlich, P. R. 1968. *The Population Bomb*. New York, Sierra Club/Ballantine Books.

Gibson-Graham, J. K. 2006. *A Postcapitalist Politics*. Minneapolis and London, University of Minnesota Press.

Goodall, H. L. Jr. 2010. *Counter-Narrative: How Progressive Academics Can Challenge Extremists and Promote Social Justice*. Walnut Creek, CA, Left Coast Press.

Guthman, J. 2011. *Weighing In: Obesity, Food Justice, and the Limits of Capitalism*. Berkeley, University of California Press.

Hamilton, C. 2011. The ethical foundations of climate engineering. http://www.schrogl.com/03ClimateGeo/DOKUMENTE/205_HAMILTON_ETHICAL_FOUNDATION_CLIMATE_ENGINEERING_2011.pdf (accessed 14 May 2017).

Harvey, D. 1996. *Social Justice and the City*. London and New York, Blackwell.

Homer-Dixon, T. 1999. *Environment, Scarcity, and Violence*. Princeton and Oxford, Princeton University Press.

Jarosz, L. 1992. Constructing the dark continent: metaphor as geographic representation of Africa, *Geografiska Annaler Series B Human Geography*, vol. 74, no. 2, pp. 105–115.

Kaplan, R. 1994. The coming anarchy, *Atlantic Monthly*, February. http://www.theatlantic.com/magazine/archive/1994/02/the-coming-anarchy/304670/ (accessed 13 June 2016).

Klein, N. 2014. *This Changes Everything*. New York, London, Toronto, Sydney, New Delhi, Simon & Schuster.

Lawson, V. 2009. Instead of radical geography, how about caring geography? *Antipode*, vol. 41, no. 1, pp. 210–213.

Merchant, C. 1980. *The Death of Nature*. New York, Harper and Row.

Nash, L. 2006. *Inescapable Ecologies: A History of Environment, Disease, and Knowledge*. Berkeley, University of California Press.

Norberg-Hodge, H. 2002. *Bringing the Food Economy Home*. New York, Zed Books.

Parenti, C. 2011. *Tropic of Chaos: Climate Change and the New Geography of Violence*. New York, Nation Books.

Plumwood, V. 2009. Nature in the active voice, *Australian Humanities Review*, vol. 46, pp. 113–129.

Plumwood, V. 1993. *Mastery of Nature*. New York, Routledge.

Watts, M. 1983. *Silent Violence: Food, Famine and Peasantry in Northern Nigeria*. Berkeley, University of California Press.

Watts, M. 2013. *Silent Violence: Food, Famine and Peasantry in Northern Nigeria*. Athens and London, University of Georgia Press.

Watts, M. 2015. All or nothing? *Human Geography*, vol. 8, no. 1, pp. 109–111.

4

Aggregate Modernities
A Critical Natural History of Contemporary Algorithms

Jake Kosek

A Critical Natural History and the Algorithm

This chapter is premised on the idea that the contemporary use of algorithms to organize large amounts of data into aggregate forms has troubling precedents within natural history and an attendant and deep set of political consequences. In many ways, the history of the use of algorithms builds on and parallels the taxonomic moves of the eighteenth and nineteenth centuries, in which a particular form of organizing knowledge radically transformed notions of the individual and the collective, the subject and populations, the atom and the totality, the element and the aggregate, among other formations that are defining features of modernity.

Importantly, though, while these new aggregates are linked to one another through the history of modernity, they have also been marked by different political understandings and moments (Le Bon 2002). Consider the horde, the mob (Hobbes 1982; Locke 1988), the population (Foucault 1978; Malthus 2007), the proletariat (Marx 1990), the public (Arendt 1958; Habermas 1989), the masses (Horkheimer and Adorno 2002; Benjamin 1969; Horkheimer 1990, the rabble (Keynes 1965) and the crowd (Freud 1981, 67–143; Le Bon 2002; Tarde 1903, 59–88). While reproducing Enlightenment-era logics of order and anarchy, each aggregate has also been animated by historically specific fears, creating different political possibilities and modes of governance. Today's

Other Geographies: The Influences Of Michael Watts, First Edition. Edited by Sharad Chari, Susanne Freidberg, Vinay Gidwani, Jesse Ribot and Wendy Wolford.

new aggregates, then, assembled from large amounts of data with the help of algorithms, both participate in this broader history and produce new consequences that warrant our attention.

While working to understand how new aggregates are made, I want to be particularly attentive to the natural history of the algorithm – that is, to the ways that the organization of the collective draws from and produces forms of nature with consequence for both human and non-human socialities. In a larger project on the political natural history of the honeybee, I trace the ways in which new human socialities have been made alongside debates in natural history. More specifically, I argue that modern understandings about the essential characteristics of naturalized collectives such as 'the poor', 'blacks' and 'savages' – embedded within threatening de-individualized masses such as the mob, the horde and the crowd – are made intelligible through interspecies histories that also make taxonomies of species, swarms and herds. These early forms reappear in modern algorithmic accounting as new aggregates, populations, interspecies assemblages and kinship formations, bearing the troubling traces of their violent natural historical pasts.

But first, to be clear, I am not claiming that algorithms are new. The algorithm's roots come from sixth-century Persian geography and astronomy (Boyer 1985; Knuth 1981). An algorithm is no more than a set of steps defining the sequence and direction of a set of operations, no more than a set of rules structuring a computational procedure (Berlinski 2000). In this way, they are like natural history's taxonomies, which are themselves no more than the rules that fix and isolate objects from their hybrid forms by defining and ordering their relations in horizontal and hierarchical forms. As with earlier taxonomies, the algorithm is a sense-making tool, a way of constituting form with the aid of delineating equations that reflect the political moment's fears and desires. These calculations and predictions, like the statistical enumerations of early state-making projects and the demarcations of colonial natural history, produced new populations and types that appeared to make manifest a preexisting order. The new aggregate forms produced by colonial knowledge claimed warrant in an underlying substrate of universal, empirical truths (see Bell 2015; Cohen 1990; Hacking 2006a). The calculations that algorithms perform start by making distinctions between objects as things and states in the world; they then bring those objects and states into relationships via new correlations that enable new imaginings and constitute new subjects, which, in turn, underwrite specific actions (Amoore and Piotukh 2016, 4; see also Callon and Muniesa 2003). Though different in their specific forms of ordering, both taxonomies and algorithms are means of structuring knowledge and thought in ways at once accurate in some contexts and deeply limiting in others.

The critiques of taxonomic classification are vast and deep, and many are relevant to the new ubiquitous forms of knowledge organization and production in an era of mass data (Boellstorff and Maurer 2015; Ritvo 1998; see also Amoore and Piotukh 2016). For my purposes, I am most focused on the ways in which the taxonomic assembling of aggregates helped produce the conditions and forms of understanding necessary to colonial projects and accumulation. As Michael Watts (2013) so clearly demonstrated in *Silent Violence*, geological ideas of desertification and degradation, as well as biological taxonomies and assessments of eco-logical risk and resilience, were vital in making natures intelligible as commodities and resources. Others have pointed out the ways in which animals and plant taxonomies have been aggregated into species by enacting boundaries, qualities and hierarchies within the living world, in the process effacing heterogeneity and difference between and within organic entities (Anderson 2007; Ritvo 1998; Schiebinger and Swan 2007; Schrader 2010). The categories inhabited by diverse entities were defined with and against other unitary, seemingly naturally defined types (Bowker and Star 1999), types that, as Foucault (1994) makes clear in the *Order of Things*, were not simply the by-product of modernity but its enabling conditions (Anderson 2007; Pratt 1992). As Foucault writes, 'one has the impression that with Linnaeus or Buffon, someone has at last taken on the task of stating something that had been visible from the beginning of time but had remained mute before...In fact, it was not an age-old inattentiveness being suddenly dissipated, but a new field of visibility being constituted in all it density' (132).

As I examine the 'density' of the process by which the algorithm's veracity is being made, tested and reproduced, I am most interested in the quotidian consequences of the algorithm's putative prescience. This requires looking not just at the residues of earlier constitutive elements of algorithms, but also its emergent features. Algorithmic representa-tions, like earlier Linnaean taxonomies, mechanize life into parts, trans-form animals into information and create digital constellations of discrete, disconnected lives, objects and environments; but it is not simply the coarse-grained sorting of individuals into types and identities that merits attention. Beyond the demographic and statistical tools of the twentieth and twenty-first centuries that linked individuals and populations through the 'average man' of the bell curve; beyond demographic profiles and types or productions of difference and varia-tion, as in the eugenics movement; algorithms also define propensities and tendencies and offer personalized recommendations that in turn shape clusters and trends (Hacking 2006a). The algorithm's effectiveness and predictive accuracy in certain moments become a claim to overall future truth. It is not merely that large amounts of data are sorted into categories – bacteria and plants, humans and animals, civil and savage, and so on – that in part condition the terms of those relationships, it is that the algorithm sorts mass volumes of data, thereby enabling new correlations that produce new aggregates and create new affinities, desires and socialities which appear more animate than earlier forms of taxonomic ordering. Because of the sheer volume and complexity of data, the algorithm creator has no way of anticipating the results of its sorting procedure and the correlations it will produce; in that sense the outcomes inhabit a realm of surprise. In short, in spite of the obvious expertise of the algorithm's creator, there is an element of genuine uncertainty that paradoxically enhances and naturalizes the credibility of its assemblages.

In other words, it is not that the algorithm is separate from the interests of its creators and moment of creation but it is also the case that what it produces is further from the creators' implicit or intentional agency, thereby making the outcomes appear to be more accurate, more objective, more true. The very distance between the results of the algorithmic procedure and the interests of the algorithm's creators, as well as the distance between those interests and the data the algorithm handles, gives the appearance of a process that is beyond the (prejudicial influences of the) social and the political. This ushers in a new digital empiricism wherein more data, more processing power and more complex self-organizing systems produce seemingly more independent intuitive insights. '"They know what we want before we do"': algorithms produce predictive ways of seeing that exceed their makers' capacities to know (Amoore 2013; Seaver 2015, 17). They predict desires and outcomes (across sectors such as insurance, casinos, national security, public health, to what shoe you or I will likely buy next) and in so doing authorize both a universalized logic and a revived empiricism.[1] To be clear: it is not a matter of rendering something more precisely or accurately, something already visible; instead it is about releasing an angle of vision that 'sees' new aggregates, that remakes the very objects of our analysis. It is this profound transformation, the remaking of the honeybee in aggregate form and its political consequences, that I am most interested in exploring here.

Of course, there are many places to observe algorithms and plenty of excellent work already being done in many fields, particularly in relationship to search engines, security and surveillance, online commerce and so on (Amoore and Piotukh 2016; Boyd and Crawford 2012; Seaver 2014). But it is not just in these domains that aggregated data and algorithms are being formulated in consequential ways; for those of us interested in the politics of modern natures, the mass collection of biological data and the algorithms that sort, order and process that data should be more central to our analysis and our politics. Furthermore, much of the work that has been done on algorithms focuses on the ways they abstract knowledge rather than the specific violent traces that result and are reproduced materially in algorithmic encounters. That is, algorithms are often written about as a form of disembodied knowledge produced by minds and machines rather than systems built upon the situated violence of individuating bodies and assembling them into aggregated collectives. My questions here are rooted in the second approach. What are the material practices by which bees are made into data and data into apiary aggregates? What contortions and transformations does the honeybee endure to become the type of aggregated data compatible with algorithmic processing? What are the broader political consequences of these ways of knowing and calculating life through algorithms? And, crucially, how does the very process of data collection work to constitute the authority of the experts it produces?

What I will ultimately argue is that the making of the bee into an algorithm has produced an abstracted and mechanized bee that is inadequate for understanding its current precarious conditions and that affords few political possibilities for consequential forms of engagement. Let me start with my own involvement in a USDA-led effort to produce predictive models of honeybee health.

The California Meeting

Over a second margarita at a Disneyland hotel's tropical forest pool during the 2015 Beekeeping Federation conference, Dennis vanEngelsdorp recounts a recent trip to a California beekeeping meeting, where he gave a presentation on the health of the region's bees. The news was not particularly good.

Nevertheless, he said, 'as I was giving the talk, I felt really good about it. I was sharing information to some of these beekeepers for the first time that might help them see the disease loads of their colonies. Real data about the health of their bees that should affect how they manage them'. He went on, 'I offered a perspective on their bees that they had not seen before'. More specifically, he had turned that information into a series of algorithms that could predict colony loss at the apiary scale to a remarkably precise (read: highly correlative) degree. It allowed beekeepers to compare their bees to others around the country. Dennis was telling the beekeepers about their futures.

But afterwards, a prominent commercial beekeeper came up to him and said it 'was the most arrogant talk he had been to in 30 years of beekeeping' (personal communication, 10 June 2015). Dennis did not go into more detail about the incident other than to shake his head and comment that 'people don't like bad news'. But I think there is something more at play in the comment. First, I know a lot of PhD entomologists, and Dennis is among the least arrogant of any I have met. I do not believe that the beekeeper's comment was about Dennis' disposition; rather, it spoke to something more significant about structural changes in the ways bees' health is known and measured. Here, a scientist who does not run bees, does not live in California and has never seen the apiary of the beekeeper can tell that beekeeper things he himself did not know or believe about the health of his hives. There is a structural change in who is the expert and who knows the bees. It is not through intimate daily interactions with hives that Dennis and his team have become leading authorities on honeybee health. The beekeeper's reaction is a commentary on this shift in the locus of authority. There are many consequences to this shift.

Of course, there is something to the way that Dennis knows bees. Dennis can pretty accurately predict the mortality of a colony or a yard based on variables such as, say, the concentrations of Varroa mites, the age structure of the hive, the time of year, the state that it is in and so on, but also what is happening to beekeepers in other places. The algorithm that organizes a national database of bee populations allows him to know a collective on a large scale and to develop probabilistic ways of seeing that can be more predictive than those of the beekeeper. But it is not enough to ask if predictions made from a national bee survey are accurate or not. Either way, this making of a collective produces effects that are consequential to beekeepers and bees.[2] How might Dennis's algorithm render bees more inert, beekeepers less political, scientists and managers more self-assured about the stability and coherence of the object of their gaze, and economists (already self-assured) more arrogant?

The Making of the Aggregate Bee

I have been meaning to get to it for weeks. But the small box marked FRAGILE LIVE BEES in red letters and the two half-litre plastic lab bottles still sit on the corner of my desk with the random detritus of important things that need to get done but have not. It is a beautiful fall day, so I take the box, bottles and a metal pan from the office kitchen and head up to my research apiary on the roof of my building at UC Berkeley.

Currently, the apiary has 10 hives. They are quite varied. Some I bought; some are from Randy Oliver's research apiaries or from friends' splits; others I acquired because I am on the call list at UC Berkeley in the event a swarm is discovered on campus. I don a veil, grab a tool that is used to remove frames from the hive and crack the lid of my favourite colony, which I have had for four years. These bees are smaller and darker than the others, and as I smoke, I feel and hear the shift in their tone. Through the opening of the bustling hive, I catch a quick glance at the queen. I take a fingerful of honey and remove my veil so I can eat it. It is one of the perks of keeping a research apiary.

These bees collect nectar from a variety of sources: the massive eucalyptus that start blooming in January, the clover on the University lawns below, the prolific jasmine and honeysuckle from the tapestry of Berkeley and Oakland gardens, the dandelions and lupine of parks and empty lots and the wild lilac, plums and blackberries from surrounding open spaces. It is here, in honey taken right from the hive, still warm from the bees and the sun, that you can best taste the elements of the floral landscape that each hive of bees has chosen and concentrated. The more you process the honey, the less present these floral distinctions become. Each hive's honey shares many of these elements but also varies depending on the way the elements are combined, the number of bees in the hive, the volume of honey produced, the concentration of mites in the hive and so on.

The yard seems mellow today, but I put my veil back on because I will be knocking the bees around a bit. I lay out the tools for the sampling I am conducting for the USDA National Survey of Bee Pests and Diseases: the plastic bottles, the metal pan from the faculty kitchen, a wash tub, a measuring cup and funnels, Ziploc bags, filter cloth, a bucket and a small shipping box, among many other items. First I prepare the shipping box, opening a petri dish with food and wetting a sponge in a test tube so the bees will have water. I remove the cover from one hive and take a central frame from the top brood box, making sure the queen is not on it, then hold it over the wash bin and knock it hard against the edge. A mass of bees falls into the bin. Scores swarm around me. I tilt the wash bin on its side, then up, and a liquid-like mass of bees pools in the lowest corner of the container. With the measuring cup, I scoop a half cup of bees and pour them down a funnel into the box. I do the same for the alcohol bottle with another half cup, put the lid on and shake it. Next I shake the remaining bees back over the hive, place the empty frame over the pan and, following carefully designed USDA protocol, firmly hit the edges of the frame against the pan three times on each side, flipping the frame in between, so that any mites, beetles and pollen fall into the pan.

I repeat this procedure with eight of my 10 hives, placing bees from all the hives into the box and then the bottle, both of which now form a composite population of my bee yard. I seal and label two boxes and bring them to the post office. The bees are pissed, and the clerk is not too happy to see us, either. No matter; I remind the kind man that this is legal. He gives me the hairy eyeball but gingerly takes the boxes and the paperwork that notes my quadrant of California, the type of beekeeper I am, the date and time I took the samples and other information about my bee yard.

These samples are not yet data, but they are the raw material and the conditions of possibility for its making. My eight composite cups of bees will be joined together with other bees from the region, constructing a quadrant that creates a sub-population that, together with the three other sections of California, will get consolidated into the state population. California's bees are in turn aggregated with those of every other state, which have also been divided into quadrants, forming a collective of the nation's bees. Combined with tens of thousands of bees from California, the West and the nation, my bees are now the geographical representatives of billions of bees from thousands of yards and complex landscape contexts tended by beekeepers with different histories and socialities. In so doing, a comparative health baseline can be established, the bees' collective condition discerned, proclamations of their health made. This amalgamated collective helps make the standard Modern National Bee.

But my bees are just beginning this journey. First, the package will arrive at the University of Maryland-managed labs of the USDA's National Honey Bee Survey (NHBS), run by Dennis vanEngelsdorp. At our meet-up by the tropical forest pool, he'd invited me to the lab, and so, on a sunny summer day, I visit the old brick building that houses the Entomology Department. The lab is its own sort of hive, full of people counting and sorting bees, going through samples, examining mite loads on the lab table, peering through microscopes, moving samples, labelling, pipetting, counting dead bees, entering data. Dennis shows me around, introduces me to the staff and then leaves me to observe and ask questions, which I do for a few days. At first, the staff is a bit puzzled, but pride in their protocol and technique leads them to explain what is going on and why. They patiently walk me through each step by which bees are processed.

First, the bees in alcohol are removed from the box they were mailed in and sorted by color so that they can be traced throughout the process. They are then passed to one of the undergraduate lab interns, who takes the lids off the containers and screws them onto a shaking machine that handles 8 to 10 bottles at any given moment. The bottles of bees are vigorously shaken for 30 minutes, just long enough to shake off the maximum number of mites and other adherents without reducing the bees to a soupy liquid mass of bee parts. At exactly 30 minutes, the liquid from the bottles is filtered out and the material in the filter is examined. The bees from the bottle are placed on a tray and sorted by another undergraduate intern, who painstakingly separates them one at a time into piles of 10. Each pile is aligned in a row within the tray to allow the intern to quickly assess whether the bees have been accurately counted. In this way, the load of mites per bee can be established and the bees can be observed for other diseases. Some of the bees are placed on a marble block and squished under a heavy marble rolling pin. The contents are then scraped off with a razor into a flask, dissolved into solvent and pipetted onto a slide in order to identify other parasites under a microscope.

Over the course of this process, measurements, quantities and observations standardize and simplify the bee into units that are then recorded in spreadsheets. Each stage is carefully delineated and marked, and samples from each bottle are carefully labelled so they can be returned to if new diseases are discovered later. The observations and other actions in this laboratory process will render different bees equivalent, making them comparable as discrete numbers, as positive or negative, healthy or diseased. Of course, calls have to be made; borderline cases have to be parsed into one dimension or the other. Each of these decisions is made among the murky differences between bees.

A similar process happens with the live bees. They are logged by time of arrival and whether or not they appear healthy, then immediately put into a freezer with a temperature of 80 degrees below zero. Of these bees, 50 are removed, placed in a mesh screen bag and ground into frozen bee dust with a mortar and pestle. Two hundred microliters of buffer are added to the bee dust, and this is divided into two 100-microliter samples of solution. A robot does the pipetting and mixing so that the sample is more consistent. The iRNA levels of different viruses are measured. As with the sample of dead bees, this one is carefully labelled, measured, filtered and examined for disease and parasite levels.

In all of these cases, the data is entered into spreadsheets along with data from samples from almost 1000 other apiaries across the United States, creating the database for the National Honey Bee Survey. This database is also combined with other information collected by Dennis's lab, organizing bees not just by region but by type of beekeeper (migratory, sideliner, hobbyist), queen health, brood patterns, hygienic behaviour and so on. (It is important to note that this data points not just to demographics but also the behaviours and preferences of the beekeeper.) All this information is then sent to a statistician, who transforms it from static columns of figures into more dynamically comparable variables, a collection of binary code 1s and 0s that is the national database of bee health. This data is separated geographically from its referent bees in Maryland and stored at a vast data centre – a different form of collective – in rural Pennsylvania.

This process of simplifying beehives from bee yard to state quadrant to region to national collective – the processing, parsing, translating and representing and storing – takes work. My bees have been segregated, mailed, shaken, cleaned, analysed, ground, pipetted, pulverized, processed and numerically rendered, individuated only to be recombined into a new digitized unity that represents the collective health of the nation's bees. This is a remarkable thing, the set of processes through which the local variations from a small sample of bees, hives and yards come to represent a vast collective. The process is so simple at one level that we tend to forget its significance, to forget what research makes: in this case, the data and archive that give form and structure to a new collective, The Modern Bee.

This national representation of The Modern Bee can be used to create means, standards and deviations, averages that will indicate the health of individual bees and hives and yards. It is not just that The Modern Bee creates a national standard but that from this standard, each individual yard's health can be judged as comparative equivalents. The creation of this equivalent sets a normative standard for the individual and the proper management of its position, health and behaviours. The collective and the individual are not two antagonistic poles in this process; they are the means of each other's making (Canguilhem 2015).

Make no mistake: this database offers a powerful perspective. In a small room adjoining the lab, I get sucked into the spreadsheets for nearly half an hour without even realizing it. The power of a spreadsheet to distil the component parts of something you care about is deeply engrossing. I want to know how my bees are made equivalent with other bees around the country in terms of mite load, brood pattern and viruses. I want to investigate the multiple variables of disease patterns in relationship to drought years, yard size and beekeeping methods. I look at my bees and see what I can determine about them, not by observing but by comparing them (or, more accurately, a small aggregated subset of them) to the collective they help constitute in an abstract and partial way.

That is what the computer facilitates. In a series of small steps – if this, then that; if that, then this – it does more than organize data, it explores relationships. The aggregate data is put into correlative relationships based on a formula that the computer helps generate. It starts backwards, searching for correlations in yards where bees have died, calculating trends in commonalities large and minute. Some are absurd – correlations between beekeeper height and mite load, age of colony and chances that the yards are in dappled shade – but slowly, as more data come in, the computer finds trends and correlations, and the algorithm is tweaked and bracketed, creating, in conversation with

Dennis, a set of findings that move beyond data points. It is this iterative aspect of the algorithm, together with large data sets, that allows data to be organized such that it 'reveals' connections beyond Dennis's individual capacities even as it yields high degrees of predictability that set the algorithm apart from common regression analysis. The algorithm starts forecasting futures, and it is these futures that Dennis presented in California.

At one level, we know this story. It is the story of calculation and the ordering of life. It is part of an industrial-scale process of making the bee into a particular collective unit that disposes it to management. I want to underline that it is not just the bee aggregate, a national bee, that is made here, but also a different way of configuring relationships. It is also worth noting that these practices and technologies constitute the scientist and his authority to speak of and for national bee health. Besides the Modern Bee, the algorithm produces something else: a stable and universal expert that has the data, comportment and authority to engineer and manage anthropogenic worlds.[3]

Conclusion

These aggregations of collectives are not separate from earlier forms. As noted in my introduction, the mob motivated Hobbesian thought, the hive Mandeville's public, the rabble Keynesian economics, the horde colonial bureaucracy, the population Malthusian social engineering, Habermasian publics and so on. These aggregates are overlapping and carry the traces of the moments of their making, have different and changing valences and bear different consequences. Tracing their natural and political histories is beyond the purview of this piece. However, in an era of Big Data and targeted consumption, healthcare and warfare, the profound consequence of these formations should make us attentive to both to the troubling residues and resonances of earlier collectives and to these new emergent aggregate forms and their effects. How are contemporary aggregations and algorithms of objects and animals made, what new taxonomies do they assemble, how do the subjects and institutions they in turn make possible bear the marks of earlier political assemblages, and how are these subjects and institutions transformed by aggregative logics?

It is also important to recognize the new political realities that algorithms produce. Older forms of collective statistics tended to be more visible. In fact, the move was to prove to people through the technique (whether bookkeeping, a census or statistical log, or a regression analysis) that what they were seeing was accurate. In this case, the algorithm is the secret syrup. The proof is not in the procedure but in the ability to predict public health trends, security threats and consumer desires. This seems significant and deeply dangerous. It is essential to understand this aspect of how elements are made into individuals, populations and trends, and the violence that marks the process of producing new collective aggregates.

At one level, we know this story. It is the story of calculation and the taxonomic ordering of life. But added to this old story is a new technical form that renders bees intelligible through the formal mechanistic ordering of the elemental parts of their biology. It is also different in that it is less about understanding the details of how those laws work and more than finding correlations that do not require an understanding of what causes this or that condition per se, and more about organizing masses of elemental bee components and sorting them in ways that are predictive of future conditions. The algorithm's success is not in its explanatory power of causation, or its detailed understanding of biological processes, but in its mechanistic ways of abstracting and aggregating the bees that makes predictions about their future. These data-based predictions (and the aggregated elements necessary for the models to work) are at the heart of the making of new apiary natures. They are at times amazing predictors of regional- and bee yard-scale decline, except for when they are not. While they appear through correlation to speak to the nature of the decline (viral presence, mite load, pesticide residues and so on) and do make visible elements not always apparent to beekeepers themselves, these ways of knowing can be profoundly narrow. Algorithms rely on the bee they produce in making their calculation, this bee is an objectified and mechanized aggregate form, and while this form sometimes offers a predictive vision of the bee, it is also a vision of the bee that is analytically blinkered and politically hobbled.

I want to underline that it is not just the bee aggregate, a national bee, that is made here, but also a different way of configuring relationships. This mass data-scale process of making the bee into a collective unit disposes it to a particular form of management, engineering and commodification. These practices and technologies constitute not just modern bee biology and its place in industrial agricultural production but also the scientist and his authority to speak of and for national bee health. In this way, the algorithm inaugurates an expert who has the data, skills and credentials to engineer and govern anthropogenic worlds. Importantly, in the case of the USDA survey, the engineer is constituted through bee natures that are not visible to the beekeeper. The figure of the engineer has a long history and is marked in particular by the struggle of the colonizer to tame, conquer and subdue the native natures of landscape and savage. The engineer's struggle happens outside and above the natures it helps produce, and these efforts at knowing and making the world make them a particular agent of history. It is this managerial transformation that I believe produced the animus towards Dennis at that California beekeeping meeting and this transformation that carries forward the violent traces of natural histories and will have new consequences of its own.

Notes

1 This is the mathematical equivalent of the God's-eye view, what Donna Haraway has called the 'God Trick' (Haraway 1986).
2 This is not new to anyone who is interested in public health, economics, sports, the histories of deviancy, incarceration, production, trade and so on. This is part of a deep history of a long shift over the past 400 years from methods of determinacy to methods of probability that marries frequencies and degrees of belief. For the geographer, it is quite similar to the map, a vast simplification of a territory that is incredibly useful because of its scale and perspective. It is powerful in part because of its limitation. Through its use and efficacy, the map is often treated as the territory itself, and the politics of selection and simplification are deeply political choices that in turn remake the physical landscape (Hacking 2006b).
3 There is, of course, a large body of literature about the structural displacement of authority with the rise of expert knowledge. What makes this case more unusual is that the conclusions are reached by the algorithms' ordering of knowledge rather than the knowledge of the expert—raising interesting questions about power-knowledge.

References

Amoore, L. 2013. *The Politics of Possibility: Risk and security beyond probability*. Durham, NC, Duke University Press.

Amoore, L. and Piotukh, V. 2016. *Algorithmic Life: Calculative devices in the age of Big Data*. London, Routledge.

Anderson, K. 2007. *Race and the Crisis of Humanism*. London, Routledge.

Arendt, H. 1958. *The Human Condition*. Chicago, IL, University of Chicago Press.

Bell, G. 2015. The secret life of Big Data, in T. Boellstorff and B. Maurer (eds), *Data, Now Bigger and Better!* Chicago, IL, Prickly Paradigm Press.

Benjamin, W. 1969. The work of art in the age of mechanical reproduction, in H. Arendt (ed.), *Illuminations: Essays and reflections*, translated by H. Zohn. New York, Schocken Books, original work published 1936.

Berlinski, D. 2000. *The Advent of the Algorithm: The idea that rules the world*. New York, Harcourt.

Boellstorff, T. and Maurer, B. (eds). 2015. *Data, Now Bigger and Better!* Chicago, IL, Prickly Paradigm Press.

Bowker, G. C. and Star, S. L. 1999. *Sorting Things Out: Classification and its consequences*. Cambridge, MA, MIT Press.

Boyd, D. and Crawford, K. 2012. Critical questions for Big Data, *Information, Communication & Society*, vol. 15, no. 5, pp. 662–679.

Boyer, C. B. 1985. *A History of Mathematics*. Princeton, NJ, Princeton University Press.

Callon, M. and Muniesa, F. 2003. Les marchés économiques comme dispositifs collectifs de calcul [Economic markets as calculative collective devices], *Réseaux*, no. 122, pp. 189–233.

Canguilhem, G. 2007. *The Normal and the Pathological*, translated by C. R. Fawcett and R. S. Cohen. New York, Zone Books, original work published 1966.

Cohen, B. S. 1990. The census, social structure and objectification in South Asia, in *An Anthropologist among the Historians and Other Essays*. New York, Oxford University Press.

Foucault, M. 1978. *The History of Sexuality*, translated by R. Hurley. New York, Pantheon Books, original work published 1976.

Foucault, M. 1994. *The Order of Things: An archaeology of the human sciences*. New York, Vintage Books, original work published 1966.

Freud, S. 1981. Group psychology and the analysis of the ego, in J Strachey (ed.), *The Standard Edition of the Complete Psychological Works of Sigmund Freud*, vol. 18. London, Hogarth Press, original work published 1921.

Habermas, J. 1989. *The Structural Transformation of the Public Sphere: An inquiry into a category of bourgeois society*, translated by T. Burger and F. Lawrence. Cambridge, MA, MIT Press, original work published 1962.

Hacking, I. 2006a. *The Emergence of Probability: A philosophical study of early ideas about probability induction and statistical inference*. Cambridge, UK, Cambridge University Press.

Hacking, I. 2006b. Making up people, *London Review of Books*, 17 August, pp. 23–26.

Haraway, D. 1986. Situated knowledges: the science question in feminism, *Feminist Studies*, vol. 14, no. 3, pp. 575–599.

Hobbes, T. 1982. *Leviathan*, ed. CB MacPherson. New York, Penguin Classics, original work published 1651.

Horkheimer, M. 1990. *Critical Theory: Selected Essays*. New York, Seabury Press.

Horkheimer, M. and Adorno, T. W. 2002. *Dialectic of Enlightenment*, translated by E. Jephcott. Stanford: Stanford University Press.

Keynes, J. M. 1965. *The General Theory of Employment, Interest and Money*. New York, Harcourt, Brace & World, original work published 1936.Knuth, D. E. 1981. Algorithms in modern mathematics and computer science, in A. P. Ershov and D. E. Knuth (eds), *Algorithms in Modern Mathematics and Computer Science: Proceedings, Urgench, Uzbek SSR; September 16–22, 1979*. New York, Springer-Verlag.

Le Bon, G. 2002. *The Crowd: A study of the popular mind*. Mineola, NY, Dover Publications, original work published 1895.

Locke, J. 1988. *An Essay Concerning Human Understanding & Two Treatises of Government*. New York, Penguin Classics, original work published 1690.

Malthus, T. R. 2007. *An Essay on the Principle of Population*. Mineola, NY, Dover Publications, original work published 1798.

Marx, K. 1990. *Capital: A critique of political economy*, vol. 1, translated by B. Fowkes. New York, Penguin Classics, original work published 1867.

Pratt, M. L. 1992. *Imperial Eyes: Travel writing and transculturation*. London, Routledge.

Ritvo, H. 1998. *The Platypus and the Mermaid, and Other Figments of the Classifying Imagination*. Cambridge, MA, Harvard University Press.

Schiebinger, L. L. and Swan, C. (eds). 2007. *Colonial Botany: Science, commerce, and politics in the early modern world*. Philadelphia, PA, University of Pennsylvania Press.

Schrader, A. 2010. Responding to *Pfiesteria piscicida* (the fish killer): phantomatic ontologies, indeterminacy, and responsibility in toxic microbiology, *Social Studies of Science*, vol. 40, no. 2, pp. 275–306.

Seaver, N. 2014. Algorithmic recommendations and synaptic functions, *Limn*, no. 2, viewed 29 August, http://www.limn.it (accessed 29 August 2014).

Seaver, N. 2015. Bastard algebra, in T. Boellstorff and B. Maurer (eds), *Data, Now Bigger and Better!* Chicago, IL, Prickly Paradigm Press.

Tarde, G. 1903. What is a society?, in *The Laws of Imitation*, translated by E. C. Parsons. New York, Henry Holt and Company.

Watts, M. J. 2013. *Silent Violence: Food, famine, and peasantry in northern Nigeria*. Athens, GA, University of Georgia Press.

5

Peanuts for Cashews? Agricultural Diversification and the Limits of Adaptability in Côte d'Ivoire

Thomas J. Bassett and Moussa Koné

The adaptive capacity much praised by geographers, anthropologists, and rural sociologists could be, and often was, undercut, eroded, or destroyed by the operations of the market

Watts, 2013, lxxviii

Introduction[1]

In the 2013 reprint of his monumental work, *Silent Violence: Food, Famine and Peasantry in Northern Nigeria*, Michael Watts writes about the 'limits of adaptability' with reference to the cultural ecological literature on African farming systems. That literature highlights the resourceful and innovative character of farmers who, in response to economic or environmental stresses, invest their time, labour and resources in a 'complex bricolage or portfolio of activities' (Scoones 2009, 172). These strategic activities involve experimenting with, developing, and selectively adopting new technologies; allocating labour to activities that minimize

Other Geographies: The Influences Of Michael Watts, First Edition. Edited by Sharad Chari, Susanne Freidberg, Vinay Gidwani, Jesse Ribot and Wendy Wolford.
© 2017 John Wiley & Sons Ltd. Published 2017 by John Wiley & Sons Ltd.

risk rather than maximize return; and engaging with markets at multiple scales (Mortimore and Adams 2001; Agrawal and Perrins 2009).

Michael Watts recognizes the value of field studies that illuminate the resourcefulness of farmers and herders faced with climatic and economic variability but offers two critiques of the adaptation concept:

> Doubtless this praise singing of the virtues of the peasantry is justified – I, too, cover similar ground – but it has little to say about the *limits* of adaptability and how the social relations of production impose constraints, rigidities, and inflexibilities. On the other hand, concerning the operations of the market and the deepening commercialization of the countryside during the crucible of neoliberalism in Nigeria, the authors are silent. (Watts 2013, lxix)

The resurgent interest in adaptation in the context of global climate change makes Michael Watts's longstanding critiques of the adaptation concept both prescient and timely (Bassett and Fogelman 2013; Head 2010; Pelling 2011; Taylor 2015). In this chapter we would like to revisit Watts's core critique of the adaptation concept, especially its theoretical underpinnings in neo-Marxist peasant studies, and discuss its relevance to current policy prescriptions in the climate change adaptation literature. We are especially interested in confronting the prescription that agricultural diversification should be in every farmer's and herder's adaptation portfolio with Watts's insistence that we consider the 'operations of the market' to assess the viability of such proposals. Following this theoretical critique of the adaptation concept, we apply it to a case of agricultural diversification in Côte d'Ivoire where farmers in the savanna region have recently added cashew cultivation to their farming portfolio. The widespread planting of cashew trees has made Côte d'Ivoire the world's number one producer of cashew nuts over the span of just 20 years. Yet, this diversification strategy has yielded uneven economic benefits to smallholder producers. In short, the structure and operations of the market have not always been favourable to cashew growers. We endeavour to show why this has been the case and conclude with a discussion on the limits and opportunities of agricultural diversification as a peasant farmer strategy to the everyday and long-term risks of economic and ecological uncertainty.

Peasants, Capital and the State: The political economy of maladaptation

In *Silent Violence*, Michael Watts dissects rural household vulnerability to drought and famine through moral economic and historical materialist approaches to peasant economy and society. He argues that the moral economy of the nineteenth century Sokoto Caliphate provided peasant households with a basic but dependable level of subsistence security thanks to multi-scale redistributive mechanisms and entitlements. The social relations of production and reproduction dramatically changed under colonial capitalism in which taxation, monetization of the economy and intensification of commodity production created new mechanisms of surplus appropriation and distribution that made peasant households increasingly vulnerable to the vagaries of climate and the market. As the conditions of household production and reproduction became increasingly mediated by the market, especially by the prices paid for cash crops and consumption goods, adverse changes in the terms of trade could lead to 'a crisis in simple reproduction' (Watts 2013, 254). Caught between falling market prices for cotton or groundnuts and higher prices of goods necessary to household reproduction such as food, seeds, animals and fertilizers, Hausa peasants were forced to either reduce consumption levels or intensify commodity production, or both. This 'simple reproduction squeeze' led to paradoxical practices such as 'cash hungry' peasants increasing the area under cotton despite falling prices (Watts 2013, 254, 350). The intensification of commodity production could also lead to land degradation (as well as intense manuring) as farmers expanded the area under cultivation and reduced fallow periods. Watts makes the compelling observation that most households were conceptually and practically prepared for the protracted droughts that hit Northern Nigeria over the twentieth century but that the social relations of production and exchange made it impossible for many to adapt to the deteriorating agro-ecological and economic conditions:

> In relation to their obvious absence of choice, partial control over the productive process, and limited flexibility and autonomy, these farming units are maladapted; that is to say, they are households constrained in their ability to respond to threats, disturbances, and perturbations. (Watts 2013, 465)

This political economic framing of the limits of adaptation represented a major advance in understanding the causal structures of hunger vulnerability in the 1980s that are highly resonate in the literature on climate change adaptation 30 years later (Bassett and Fogelman 2013). Indeed, the continued relevance of Watts' historical materialist critique of the adaptation concept in the 1970s and 1980s (Adaptation 1.0), is apparent in his 2015 contribution to the *Routledge Handbook of Political Ecology* in which he roundly critiques the concept's recent revival in the climate change adaptation and resiliency literatures (Adaptation 2.0) (Watts 2015). A salient thread in Watts' critique of the

1970s and 2010s versions of adaptation is the common conception of adaptation as 'purposive adjustment' to external perturbations in which the broader political economy is overlooked. In the next section we take up Watts's refrain about the need to consider the 'operations of the market' in the climate change adaptation literature. We do so with reference to that literature's emphasis on agricultural diversification as a viable adaptive strategy to adverse environmental and economic conditions. The question guiding our inquiry is under what conditions do farmer initiatives to diversify crop production lead to increased incomes and options?

Expanding the Portfolio: Agricultural diversification and adaptive capacity

The most recent report of the Intergovernmental Panel on Climate Change (IPCC) titled *Climate Change 2014: Impacts, Adaptation, and Vulnerability*, refers repeatedly to livelihood diversification as a promising adaptation option.[2] In rural areas livelihood diversification refers to expanding both on-farm (adopting hardier crops; crop diversification) and off-farm (rural enterprises, migration) activities that will spread risk and increase the incomes of rural households. More stable and higher incomes will theoretically increase the options of rural households faced with climate variability. The authors of the regional chapter on Africa acknowledge the resourceful and innovative character of the continent's farmers:

> Inherent adaptation-related strengths in Africa include the continent's wealth in natural resources, well-developed social networks, and longstanding traditional mechanisms of managing variability through, for example, crop and livelihood diversification, migration, and small-scale enterprises, all of which are underpinned by local or indigenous knowledge systems for sustainable resource management. (Niang et al. 2014, 1226)

Despite these adaptive capabilities, Africa is portrayed as having an 'adaptation deficit' as a result of multiple constraints or 'barriers' that limit adjustments to the projected climate changes. The checklist of barriers is extensive. It includes institutional, financial, political, biophysical, technological, cognitive, discursive and infrastructural barriers. The authors of the Africa chapter discuss these constraints one-by-one in a non-theorized manner. The obstacles are many and seemingly insurmountable for Africa's rural poor. Only governments, the private sector and NGOs seem capable of rescuing Africa through so-called 'low regrets policies' and development interventions. In summary and as Marcus Taylor argues, adaptation is constructed by the IPCC as a 'new object of development' (Taylor 2015, xi–xii).[3] Absent from the IPCC's framing of adaptation is any mention of the broader political economy that is implicated in both climate change processes and the causal structures of vulnerability (Ribot 2011, Pelling 2011; Taylor 2015). In this sense, the 2014 report follows the previous assessments in conceptualizing adaptation as an adjustment process to prevailing political-ecological systems.

Given the IPCC's emphasis on economic and livelihood diversification as an adaptation strategy, one would expect to find a discussion of commodity markets and how they work or do not work to generate income for the rural poor. There is no mention in the IPCC's Fifth Assessment of market structure, trends in terms of trade, and the challenges that farmers might face in diversifying production. In short, 'market barriers' are not part of the analysis. There is a brief discussion in one adaptation chapter on the relationship between global coffee prices and household vulnerability and adaptive capacity (Klein et al. 2014, 918). This example points to an important market-society relation that warrants further attention. In the chapter titled 'The Economics of Adaptation' (Chambwera et al. 2014), the authors have little to say about price formation of primary agricultural commodities and its effects on the adaptive capacity of nearly half of the world's population. In the chapter devoted to rural areas, there is a brief and inconclusive review of the literature on the effects of increased market participation on the adaptive capacity of 'small famers and poor people'. In some cases, market integration increases vulnerability by 'accelerating socioeconomic stratification and reducing crop diversity' (Dasgupta et al. 2014, 634). But in other cases, distance from markets increases vulnerability. The authors conclude that 'each case needs to be analyzed within its complexity, considering interactions among all the factors that can affect vulnerability' (Dasgupta et al. 2014, 634).

The consideration of 'all the factors that can affect vulnerability' leads the authors of the various IPCC chapters to chain together longs lists of variables that could affect the adaptive capacity of the rural and urban poor. This style of writing may be attributed to the state-of-the-art structure of the IPCC report and what Forsyth refers to as the type of knowledge characteristic of boundary organizations that shapes science and policy 'through practices such as problem closure and framing' (Forsyth 2003, 142). In the end, the IPCC report leaves the reader with little understanding of how the operations of the market might affect the viability of agricultural diversification as an adaptive strategy. In the following sections we address this broader question with reference to crop diversification in Côte d'Ivoire where peasant farmers, particularly men, have added cashew nut cultivation to their farming systems.

The Expansion of Cashew Cultivation in Côte d'Ivoire

Over the past 20 years, Côte d'Ivoire has risen from being an insignificant producer of cashew nuts to the world's largest producer. In 2015, it produced more than 700 000 tons of raw cashew nuts (RCN),[4] surpassing India (650 000 T) and Vietnam (325 000 T), the second and third largest cashew nut producers in the world (Abidjan.net 15 February 2016). The remarkable thing about Côte d'Ivoire's rise to the top is that it is a classic bottom-up development story. For the past two decades, peasant farmers in the savanna region have been planting cashew trees with the goal of diversifying their agricultural systems which have been up until now centred on cotton. They typically plant cashew seedlings in their peanut fields at the end of 8–10-year rotation of food crops and cotton. The trees bear fruit in the fourth year after planting. The case of Katiali in the Korhogo region of north central Côte d'Ivoire offers insights into the process of agricultural diversification.

Figure 5.1 illustrates the expansion of cashews in the farming systems of 40 households followed by Bassett and more recently by Koné over the past 35 years in the village of Katiali.[5] It shows the percentage of the area cultivated in different crops and crop combinations by these households for selected years. For the sake of legibility, we have aggregated the various crops and combinations into seven major groups in Figure 5.2. The major trend that pops out in this graph is the relative decline in cotton and increase in the cashew nut area. To understand the dynamics of this change and assess its benefits to Katiali's farmers, we need to examine the conditions of production and exchange of these two crops.

Production and Market Relations of Cotton and Cashews

Cotton and cashews are complementary crops in that they do not compete with each other for labour time. Cotton is a labour-intensive row crop that is hand harvested during the early dry season months of November and December. Cashews on the other hand are perennial tree crops that are harvested during the middle dry season months of February and March. The contrasts between the two crops are striking in terms of their labour demands, capital requirements, market relations and levels of indebtedness. Each of these areas is discussed separately below. Table 5.1 summarizes the main differences.

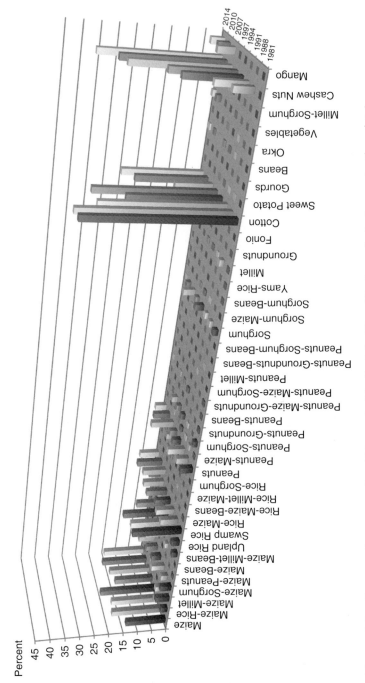

Figure 5.1 Crops and crop area for 40 households, Katiali, Côte d'Ivoire, selected years (author data).

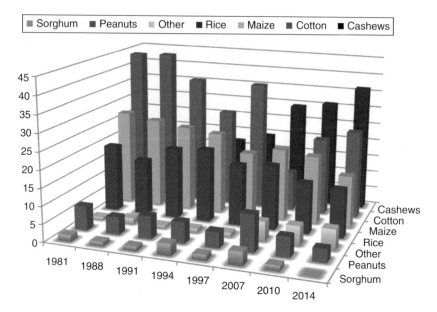

Figure 5.2   Combined crop area for 40 households, Katiali, Côte d'Ivoire, selected years (author data).

In addition to being labour intensive, cotton requires considerable capital investment for chemical fertilizers and pesticides. Table 5.2 indicates the agro-input costs for cultivating one hectare of cotton in 2011.

At an exchange rate of 500 CFA francs per US dollar, the agro-input costs of cultivating one hectare of cotton amounted to $350 in 2010/11.[6] These costs are deducted from farmers' earnings by cotton companies who provided inputs on credit to farmers at the beginning of the growing season. Agro-input loans typically amount to 45% of farmers' gross earnings. In 2010/11, they accounted for close to 50% of the gross revenues of the cotton growers in the 40-household sample. In addition to reimbursing cotton companies for that year's agro-inputs, farmers also paid down outstanding debts incurred in previous years.[7] In contrast, cashew growers spend comparatively little in maintaining their orchards. Some off-farm labour is employed to weed orchards and to harvest nuts. Workers are often paid in-kind (e.g. raw cashew nuts) for their labour time. A broad spectrum of cashew growers also use herbicides to control weeds.

Cotton grower incomes vary with the price of agro-inputs, yields and the producer price for seed cotton. In 2010/11, the producer price was set at 210 FCFA/kg and yields averaged 1085 kgs/ha for the sample of 40 households.[8] That year, the net income from a hectare of cotton amounted to a meagre $170. Cotton incomes improved the following

Table 5.1 Production and market relations for cotton and cashew nuts.

Characteristic/Crop	Cotton	Cashew
Labor demands	High	Low
Capital investment	High	Low
Land rights	Low	High
Market structure	Oligopsonistic	Competitive
Prices	Fixed	Vary
Indebtedness	High	Low
Women's access	Low	Low

Source: Farm surveys, Katiali, 1981–2015.

Table 5.2 Agro-chemical input costs for cultivating one hectare of cotton, Katiali, Côte d'Ivoire, 2010/11.

Agro-chemical inputs	Total cost per hectare
4 sacks of NPK fertilizer	54,000 FCFA
1 sack of urea	12,500 FCFA
4 litres/ha of insecticide at 4333 FCFA/ litre x 6 sprayings	106,392 FCFA
Total agro-chemical costs/ha	172,892 FCFA

Source: Farming System Survey 2012, Katiali.

year when the price was set at 265 FCFA/kg and yields attained 1160 kgs/ha. After credit repayments, the average earnings for a hectare of cotton in 2011/12 came to $380.[9]

In contrast, cashew growers earn less per hectare but expend far less labour, incur far fewer costs and do not fall into debt. Cashew orchard yields are low, ranging between 200 to 300 kgs/hectare (GIZ 2010). Unshelled raw cashew nut (RCN) prices vary monthly, yearly and by region. However, if we take a three-year (2011–2013) average of 191 FCFA/kg for the Savanna region,[10] and estimate yields at 250 kgs/ha, the average cashew grower earns about 50 000 CFA/ha or $110/ha.

The producer price for unshelled raw cashew nuts is set annually in Côte d'Ivoire by the cotton and cashew regulatory authority.[11] Between 2005 and 2013, it was called an 'indicative price', meaning that it was a price that buyers should respect (*L'Anacardier*, September–October 2012, 9). But traders rarely respected these prices. Market reports throughout this period note that RCN prices were well below the official price. In Katiali, local buyers paid farmers 35–50 FCFA/kg of RCN in 2008 when the official market price was 150 FCFA. The year before, merchants bought nuts for the abysmal price of 15 FCFA/kg despite an official price of 150 FCFA. When farmers receive 'peanuts' for cashews, their ability to strengthen their adaptive capacity to adverse economic and climate conditions is highly constrained.

The 2014 reform of the cashew sector introduced a major change to producer prices. The official price for RCN is now a guaranteed floor price (RC1 2013). That is, merchants must now pay producers at least the floor price for their crop. The purchase price can go higher as it did in 2015 and 2016 when world demand and market prices were high. It remains to be seen if the Cotton and Cashew Council will enforce the floor price when market prices fall below that threshold. When prices are high and the state performs its regulatory role then the benefits of diversification can be realized.

In contrast to the competitive nature of the cashew economy in which there are multiple buyers and price competition, the cotton economy is marked by its oligopsonistic structure and fixed prices. Just a few companies dominate the cotton market in Côte d'Ivoire. Their control has historically been facilitated by the division of the cotton growing areas into geographical zones in which a single company has the exclusive right to buy cotton. Cotton prices are the same throughout the cotton growing areas and do not change during the year.

In summary, the structure of markets matters in any evaluation of agricultural diversification as an 'adaptation option'. The oligopsonistic structure of the cotton economy gives farmers few options to improve their income and livelihoods. Combined with the capital- *and* labour-intensive nature of cotton growing, farmers are more often than not between the rock of high production costs and the hard place of low producer prices. These are the conditions that are ripe for the simple reproduction squeeze that Michael Watts theorized as leading to the impoverishment of Africa's peasantries. These difficult circumstances also induce rural producers to diversify production in order to improve their incomes and livelihoods.

Market Structures and Producer Prices in the Cashew Economy

The benefits of agricultural diversification are inextricably tied to commodity relations and power struggles among producers, merchants, cashew processors and the state over the distribution of value. The ability to benefit from cashew growing thus hinges on peasant farmers' capacity to capture the most value generated in the production and exchange processes. In the example below, producer prices serve as a proxy measure of the value captured by cashew growers.

To examine the relationship between market structures and producer prices in Côte d'Ivoire's cashew economy, we conducted a comparative study in three communities located in the central and northern cashew-growing regions of the country: Sépingo in the Bondoukou (*Gontougo*) region, Krofoinssou in the Bouaké (*Gbéké*) region and Katiali in the Korhogo (*Poro*) region. The sites were selected for their contrasting market structures and for regional price differences.[12] We also interviewed the principal actors in the cashew economy (rural traders, regional merchants, nut processing factories, exporters and regulatory authorities) to gain their perspectives on the organization of markets and price formation. Table 5.3 outlines the major characteristics of markets at these three sites. The remainder of this section summarizes the results of this comparative study of production relations, markets and producer prices.

The relations of production and value capture in Côte d'Ivoire's cashew economy take two principal forms: simple commodity production and taxation. There are three forms of commodity relations: (1) the individual producer and multiple intermediaries relation, in which merchant's capital dominates; (2) the informal marketing group relation, in which industrial capital ties producers into distinctive commodity relations; and (3) the cooperative marketing relation involving direct sales to local processors and exporters by organized groups of producers (Figure 5.3). In all three cases, cashew growers are independent peasant producers engaged in simple commodity production (Bernstein 1979). They all possess individual orchards whose sizes vary between three and five hectares, and grow food crops for household consumption. In addition to simple commodity production, state taxation is a second

Table 5.3 Comparison of production relations, market structures and producer prices in three cashew-growing areas of Côte d'Ivoire.

Market/Price Relation	Katiali	Krofoinsou	Sépingo
Market structure	Trader dominated	Informal contract	Cooperative
Production relations	Merchant's capital	Industrial capital	Merchant's capital
Number of intermediaries	Many	None	None to one
Market knowledge	Poor	Fair	Good
Negotiating power	Weak	Weak	Fair
Receives floor price or better	Rarely	Sometimes	Often

Source: Farming System Surveys 2013, Katiali, Krofoinsou, and Sépingo.

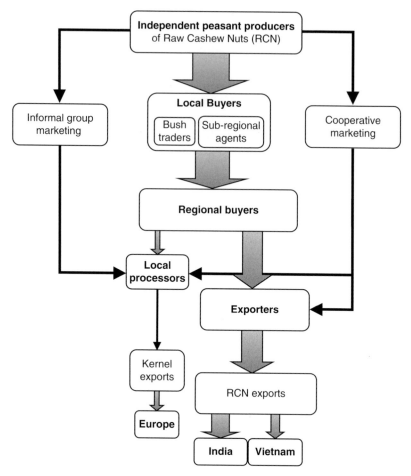

Figure 5.3   Market structure for unshelled raw cashew nuts in Côte d'Ivoire.

mechanism of surplus transfer from peasant producers to non-producers (Deere and de Janvry 1979). We discuss each of these patterns of surplus extraction in the remainder of this section.

Peasant Commodity Production and Exchange

The most common form of production and exchange is the unorganized individual producer who sells directly to merchants in his/her community (Figure 5.3). This is the relation that characterizes cashew production and marketing in Katiali. There local buyers use their own capital or are pre-financed by regional buyers and their agents (sub-regional buyer) to purchase raw cashew nuts. Local buyers often work with a phalanx of bush traders who travel to neighbouring villages and markets to buy RCN. Regional buyers are typically financed by large exporting firms. In sum, this dense cashew-trading network produces many market intermediaries who operate between individual cashew growers and exporters. Each buyer has a profit margin that cumulatively drives down the producer price.

The second form of commodity relations linking independent peasant producers with cashew markets is through informal group marketing. This relationship links individual growers organized into informal groups by cashew-processing factories. The case of Krofoinsou exemplifies this market structure. Cashew growers there sell their crop directly to OLAM International's processing plant in Bouaké. In compensation for provisioning the OLAM factory, the group is promised an attractive market price plus a marketing commission of 12.5 FCFA/kg for assembling nut shipments to the OLAM plant. Selling directly to the processing factory offers many potential advantages to Krofouinsou's cashew growers. First, by eliminating intermediaries from the commodity chain, farmers should be able to capture a larger share of the value of their product (Konan and Ricau 2010). Second, the prospect of higher and/or more stable prices should bring some security to households participating in the informal contract supply chain. And third, working closely with a leading global trader in food should impart knowledge about product quality. Some knowledge transfer did occur in the context of the Cashew Supply Chain Management Project that was co-financed by OLAM and GIZ, the German bilateral aid agency. The project emphasized best practices for the cultivation and post-harvest management of cashew nuts on the assumption that higher quality would lead to higher producer prices. But the anticipated market benefits were not realized. OLAM stopped buying from the Krofoinsou group in the middle of the 2013 harvest. When it finally informed the group it would buy again, OLAM dropped its buying price to below the official floor price (Core 2013).

The third commodity chain structure involves producer cooperatives that sell RCN directly to exporters and processing plants. The case of Sépingo is illustrative. Cashew growers in that community only sell through the cooperative. Cooperative staff measure the kernel outturn ratio (KOR) of RCN, an industry standard metric of raw cashew nut quality. The staff contacts export companies and processing plants with offers to sell a certain number of tons of RCN at a specific KOR. Its knowledge of nut quality combined with its ability to supply large quantities of nuts puts the Sépingo cooperative in a strong price negotiating position. Not surprisingly, Sépingo growers consistently receive a market price that is above the official price.

In sum, cashew growers in Côte d'Ivoire share the similar social category of simple commodity producers. They are all independent peasant farmers who produce for the market as well as for subsistence. What distinguishes them is their different relationships with agrarian capitalism. The majority of producers are entangled in long chains of accumulation in which merchants' capital dominates (e.g. Katiali). As the number of intermediaries declines and producers become better organized, some cashew growers are able to capture a larger share of the value of raw cashew nuts (e.g. Sépingo). The case of Krofoinsou indicates that conditions in which the anticipated monetary gains of short supply chains for producers are not realized. Rather than capture a larger share of the value added in short commodity chains, farmers became captive to OLAM's supply chain scheme. The point is that such differences in the social relations of production and exchange are key considerations in assessing how agricultural diversification might or might not contribute to the adaptive capacity of peasant farmers.

State Taxation

Export taxes and special levies on raw cashew nut exports represent a second arena in which value is transferred from peasant producers to non-producers. This transfer takes place through the cashew price setting mechanism in which a half-dozen charges are made at the expense of farmer incomes. The state imposes an export tax of 10 FCFA for each kilogram of RCN leaving the port. In addition, special fees ranging from 1.5 to 5.33 FCFA/kg are set for (1) financing RCN quality control at the port (1.5 FCFA); (2) funding the Cotton and Cashew Council (3 FCFA); (3) providing jute sacks to producers (5.33 FCFA); (4) investing in the cashew stabilization fund (3.5 FCFA); (5) road maintenance (2 FCFA); and (6) for research and development (1.75 FCFA/kg).

Over the three-year period (2013–2015), the export tax amounted to $11.6 million per year. The three-year average for the special fees came to $15.7 million. In total, the combination of export taxes and special levies on raw cashew nut exports entails the transfer of more than $27 million from farmers to the state each year. The question is whether these appropriations contribute to their collective needs for better rural roads, higher yielding cashew trees and reduced price risk during market slumps.

Conclusion

In comparison to its previous reports on global climate change, the Fifth IPCC assessment devotes considerable attention to the issue of adaptation to existing and future climate change. Four chapters are dedicated to the topic and at least a half-dozen more discuss it at length. Despite its salience in the report, the discussion of the adaptation concept is strikingly familiar. Like the previous volumes, adaptation is theorized as simply adjusting to external stimuli. There is comparatively little attention to the causal structures of vulnerability. This chapter seeks to advance the climate change adaptation literature by bringing systematic attention to the broader political economy where vulnerability and opportunity can be more productively discussed. As Michael Watts has indefatigably argued, we need to examine the 'operations of the market' and how 'the social relations of production impose constraints, rigidities and inflexibilities' (Watts 2013, lxxvii; lxix). This chapter's focus on agricultural diversification as an 'adaptation option' takes Watts' call seriously. The case study of Côte d'Ivoire's cashew boom allows us to draw some tentative conclusions about the ability of Africa's rural poor to benefit from agricultural diversification.

Our first observation centres on the agronomic characteristics of agricultural diversification. The differences between cotton and cashews could not be greater in terms of their labour and capital requirements and growing conditions (annual row crop; perennial tree crop). Farmers also highlighted these differences in the 2015 farming systems survey when discussing why cashews are a better cash crop than cotton. The 'fit' of cashews within local farming systems is key to the viability of this example of crop diversification.

A second insight revolves around the social relations of production and the distribution of value of the commodities sold by peasant farmers. The ability of Sépingo's farmers to reap a higher price for their crop stems from their stronger negotiating power vis-à vis merchants and processing firms. In contrast, Katiali's cashew growers are price takers in relation to the plethora of intermediaries operating between them and the port of Abidjan. The failure of Krofoinsou's farmers to capture a larger share of the value of their crop sold directly to the OLAM factory reflects their limited power in this market structure. In all instances, farmers are independent peasant producers. It is their integration into different market structures and production relations that determines their ability to benefit from diversification.

Finally, the cashew case study shows that the state plays a key role in determining the viability of crop diversification as an adaptation option. State regulatory policies like the cashew floor price can ensure that farmers receive a higher share of the value of their crop than they would obtain in an unregulated market. The floor price can also potentially add some stability in rural markets that could help to reduce the rural poor's vulnerability to debt, to food insecurity, and other immediate and long-term concerns. On the other hand, the state's taxation of cashew exports for revenue generation and financing of the Cotton and Cashew Council and other state activities represents a redistribution of value from producers to non-producers. Whether this taxation is justified or not remains to be seen. If the Council sets and enforces a fair floor price then it might legitimate its existence, and farmers will not receive 'peanuts' for cashews as in previous years. In summary, we need to disentangle the relations of production and the operations of the market to gain a clearer understanding of the opportunities and limitations of the adaptive strategies of peasant farmers in this era of global climate change.

Notes

1 We thank the National Geographical Society for a field research grant (#9103-12) that made this study possible.

2 Agricultural diversification is discussed as an adaptation strategy in at least six chapters (7, 9, 13, 14, 16 and 22) in the IPCC's Fifth Assessment.

3 As an object of development, Taylor describes adaptation as 'a normative goal and framework within which practical interventions are planned, organized, and legitimized' (Taylor 2015, xi–xii).

4 RCN are nuts in their shell. When the nuts are removed from the shell, they are called kernels.

5 This is the same community where Bassett did his doctoral dissertation research in 1981–82 under the guidance of Michael Watts. While a graduate student at the University of Illinois at Urbana-Champaign, Koné conducted his doctoral dissertation research in Katiali in 2006–08 under Bassett's supervision. Koné administered the 2014/15 farming system survey whose data appear in this chart.

6 Cotton growers also employ off-farm labour for weeding and harvesting the cotton crop, which amounts, on average, to an additional $50 per hectare.

7 In 2010/11 farmers were still paying off debts incurred from the 2006/07 and 2007/08 seasons when 78% and 66% of farmers respectively in the household sample fell into debt.

8 Producer prices for cotton in Côte d'Ivoire are set annually through a complicated price-setting mechanism that is biased towards cotton-ginning companies (Bassett 2013).

9 These figures are based on data collected from the sample of 40 households. In 2010, 21 households grew cotton. That number increased to 30 in 2011/12.

10 The data are derived from the weekly cashew market newsletter, *N'Kalo*, published by Rongead, a French NGO. http://www.rongead.org/N-KALO-Project-Ivory-Coast.html (accessed 15 May 2017).

11 This institution was called ARECA (Autorité de Regulation du Coton et de l'Anacarde) between 2002, the year it was created, and 2013, the year of the cotton and cashew reform. Since, 2013, the authority is known as the Cotton and Cashew Council (Le Conseil de Coton et de l'Anacarde) (CCA 2015).

12 Information on regional prices is published weekly during the marketing season by the French NGO Rongead in its newsletter, *N'kalo*. http://nkalo.com (accessed 15 May 2017).

References

Agrawal, A. and Perrin, N. 2009. Climate adaptation, local institutions and rural Livelihoods, in W. N. Adger, I. Lorenzoni and K. O'Brien (eds), *Adapting to Climate Change: Thresholds, Values, Governance.* Cambridge, UK: Cambridge University Press, pp. 350–367.

Bassett, T. 2013. Capturing the margins: world market prices and cotton farmer incomes in West Africa, *World Development*, vol. 59, pp. 408–421.

Bassett T. and Fogelman, C. 2013. Déjà vu or something new? The adaptation concept in the climate change literature, *Geoforum*, vol. 48, pp. 42–53

Bernstein, H. 1979. African peasantries: a theoretical framework. *Journal of Peasant Studies*, vol. 6, no. 4, pp. 421–443.

Chambwera, M. et al. 2014. Economics of adaptation, in *Climate Change 2014: Impacts, Adaptation, and Vulnerability.* Part A: Global and Sectoral Aspects. Contribution of Working Group II to the Fifth Assessment Report of the Intergovernmental Panel on Climate Change. Cambridge and New York, Cambridge University Press, pp. 945–977.

Conseil du Coton et de l'Aanacarde (CCA). 2015. *Recuil de texts applicable aux filières coton et anacarde.* Abidjan, CCA.

Core, E. 2013. Evaluation du projet PPP-GIZ-OLAM 'Gestion de la Chaine d'Approvisionnement de l'Anacarde'. Rapport d'Evaluation. No du contrat: 83158661 CO 110 2013.

Dasgupta, P. et al. 2014. Rural areas, in *Climate Change 2014: Impacts, Adaptation, and Vulnerability.* Part A: Global and Sectoral Aspects. Contribution of Working Group II to the Fifth Assessment Report of the Intergovernmental Panel on Climate Change. Cambridge and New York, Cambridge University Press, pp. 613–657.

Deere C. D. and de Janvry, A. 1979. A conceptual framework for the empirical analysis of peasants, *American Journal of Agricultural Economics*, vol. 61, no. 4, pp. 602–611.

Forsyth, T. 2003. *Critical Political Ecology: The politics of environmental science.* New York, Routledge.

GIZ. 2010. *Amélioration variétale.* Abidjan, GIZ.

Head, L. 2010. Cultural ecology: adaptation – retrofitting a concept? *Progress in Human Geography*, vol. 34, no. 2, 234–242.

Klein, R. J. T. et al. 2014. Adaptation opportunities, constraints, and limits, in *Climate Change 2014: Impacts, Adaptation, and Vulnerability.* Part A: Global and Sectoral Aspects. Contribution of Working Group II to the Fifth Assessment Report of the Intergovernmental Panel on Climate. Cambridge and New York, Cambridge University Press, pp. 899–943.

Konan, C. and Ricau, P. 2010. *La filière anacarde en Côte d'Ivoire. Acteurs et organisiation.* Consultant report. Lyon, France.

Mortimore, M. J. and Adams, W. M. 2001. Farmer adaptation, change and crisis in the Sahel, *Global Environmental Change*, vol. 11, pp. 49–57.

Niang, I. et al. 2014. Africa, in *Climate Change 2014: Impacts, Adaptation, and Vulnerability.* Part B: Regional Aspects. Contribution of Working Group II to the Fifth Assessment Report of the Intergovernmental Panel on

Climate Change. Cambridge and New York, Cambridge University Press, pp. 1199–1265.

N'Kalo. 2015. *Cashew Market Bulletin* No. 158, 9 June 2105.

Pelling, M. 2011. *Adaptation to Climate Change: From resilience to transformation*. New York, Routledge.

République de Côte d'Ivoire (RCI) 2013. *Document d'Operationnalité de la Réforme des Filières Coton et Anacarde*. Abidjan, Ministère de l'Agriculture.

Ribot, J. 2011. Vulnerability before adaptation: toward transformative climate action, *Global Environmental Change*, vol. 21, no. 4, pp. 1160–1162.

Scoones, I. 2009. Livelihoods perspectives and rural development, *Journal of Peasant Studies*, vol. 36, no. 1, pp. 171–196.

Taylor, M. 2015. *The Political Ecology of Climate Change Adaptation*. London and New York, Routledge and Earthscan.

Watts, M. 2013. [1983] *Silent Violence: Food, famine, and peasantry in northern Nigeria*. Athens and London, University of Georgia Press.

Watts, M. 2015. 'Now and then: the origins of political ecology and the rebirth of adaptation as a form of thought', in T. Perrault, G. Bridge and J. McCarthy (eds), *Handbook of Political Ecology*. London, Routledge, pp. 19–50.

6

Life Itself Under Contract

A Biopolitics of Partnerships and Chemical Risk in California's Strawberry Industry

Julie Guthman

In 'Life under Contract', Michael Watts (1993, 33) famously wrote that certain kinds of farming contracts make the farmer 'little more than a propertied labourer, a hired labourer on his or her own land'. This was one of several pieces in which he invoked the agrarian question to understand how different relations of production shape the conditions of agrarian livelihoods. In conversation with others writing in the agrarian political economy tradition, he argued that contract relations seem to prevail wherever capital seeks both maximum control and minimal risk – and especially risk posed by the vagaries of nature, including perishability, weather, pests and disease. By leaving those risks to ever more marginal producers, contract farming helps explain why peasants persist in an ever more capital-dominant world.

Such foundational concepts in agrarian political economy remain compelling, yet I am nonetheless struck by this conceptualization of biological risk as primarily livelihood risk: the risk, that is, of failed crops or sick animals is that a contract farmer potentially loses income or even the contract itself. Building off Watts' insights and drawing on new theoretical developments, I suggest that contract farming devolves biopolitical responsibility, as well as political economic risks to growers, who face difficult choices about making life live. As Foucault (1985, 1997, 2007) often averred, making life live depends on determining what life is worthy, and then either putting the rest to death or letting it die. Foucault had little to say about agriculture – he was primarily

Other Geographies: The Influences Of Michael Watts, First Edition. Edited by Sharad Chari, Susanne Freidberg, Vinay Gidwani, Jesse Ribot and Wendy Wolford.
© 2017 John Wiley & Sons Ltd. Published 2017 by John Wiley & Sons Ltd.

concerned with the emergence of new political knowledges and disciplines that made it possible to secure the life of the population. But his recurring comments on sorting surely apply to agriculture, which necessarily involves killing life to make life. Therefore, mechanisms that prescribe agricultural practices, such as contracts, are implicitly prescriptions of whose and what lives count.

This intervention draws from my research on fumigant use in California's strawberry industry, where contract production predominates. I am most interested in the contracts brand name shippers make through less visible intermediaries with former farmworkers cum ranch managers, whom they designate as 'partners'. These partners assume enormous debt to lease equipment and land, purchase inputs and pest control services, hire labour and purchase marketing materials. Managing partners provide and are paid for many of these services, arguably as rents, and in addition earn proceeds proportionate to the share of capital they invest. The arrangement is tantamount to what some have called debt peonage on leased land (Schlosser 1995; Wells 1996, and many critics from within the industry). These farmers are particularly inclined to use soil fumigants and other agro-chemicals because they farm land where strawberry plants are highly vulnerable to certain soil pathogens. Importantly, these fumigants are highly toxic to workers, neighbouring communities and farmers themselves, but not to consumers. Thus farmers must assume the responsibility to manage not only plant life – and pest death – but also the human lives in the vicinity of strawberry production.

Watts and the Agrarian Question

Given Watts' influence on my own and others' scholarship, I want to begin with his theorizations of the agrarian question and contract farming more specifically. On the former, Watts grappled with how non-capitalist forms of agrarian production persisted within capitalism. Much of his work on this subject stemmed from his research on peasantries and famine in Sub-Saharan Africa. Characteristically, it came during a tremendously generative time in agro-food studies (newer arrivals in this field would do well to revisit some of the classics of late seventies, eighties and early nineties). But while others declared agriculture thoroughly industrialized, globalized or in the midst of epochal regime shifts, Watts insisted on geographical specificity. He refused the stark lines drawn between agrarian forms, instead tracing out overlaps and articulations. He also took seriously how the character of commodities figured in these hybrid forms.

His approach was largely inspired by his return to the classics. Like some of his contemporaries in agro-food studies (e.g. Harriet Friedmann, Susan Mann), he read Chayanov, Lenin and Marx for insights about peasant differentiation and agrarian transitions. He was especially taken with Kautsky (1988), who predicted in 1899 that farms themselves would mostly remain as non-capitalist enterprises even as capitalism took hold around them. Introducing their *Globalising Food* volume, Goodman and Watts (1997, 9) wrote:

> Kautsky concluded that industry was the motor of agricultural develop-
> ment – or more properly, agro-industrial capital was – but that the peculi-
> arities of agriculture, its biological character and rhythms coupled with
> the capacity for family farms to survive through self-exploitation (i.e.,
> working longer and harder to in effect depress 'wage levels'), might hinder
> some tendencies, namely, the development of classical agrarian capitalism.
> Indeed agro-industry – which Kautsky saw in the increasing application of
> science, technology and capital to the food processing, farm input and
> farm finance systems – might prefer a non-capitalist farm sector.

Contract farming seemed to exemplify the agrarian forms Kautsky fore-saw. Vertically integrated companies would provide and often dictate all necessary inputs to the farmer; the farmer would take responsibility for the grow-out portions of production and their attendant risks, and then the company would process and sell the final product, whether plant or animal. The terms appropriationism and substitutionism, coined by Goodman, Sorj and Wilkinson (1987) referred to the same phenomena, although they extended Kautsky's theorizations to suggest that capital actively commodified previously 'natural' aspects of agrarian produc-tion and then sold them back to farmers in the form of seeds, machinery and pest control agents. Watts (1993) also differentiated marketing from production contracts, noting that the latter's regulation of practices, product quality and credit stripped farmers of nearly all autonomy, while burdening them with debt that furthered their subordination.

For Watts (1993), contract farming signified a major shift away from plantation production, not unlike historic shifts from direct to indirect forms of colonial control. Therefore, consumer demand for foods with qualities that created variable incentives and diseconomies of scale (25), such as fresh fruits and vegetables (FFV), was one of the driving forces of contract farming. Resistant to mechanization, FFV production had to meet the dual challenge of keeping labour costs as low as possible while ensuring the quality of fragile and highly perishable crops (Goodman and Watts 1997). It appeared that small, family-sized units of produc-tion could best address these dual challenges while still allowing capital to concentrate outside the unit of production (Watts 1993).[1]

While much of 'Life under Contract' focused on FFV production in the global South, elsewhere Watts showed how contract farming plays out within other sectors, within the United States. Along with William Boyd (1997), for example, he linked the US chicken industry's move to the US south to the region's abundance of poor farmers with poor land. Boyd and Watts paid careful attention to the breeding programmes that increased chicken's growth rates but also their vulnerability to disease. The liveliness of chickens was central to this analysis – what Watts referred to as the biological, or organic (Goodman and Watts 1997). Chicken farmers had to carefully monitor their flocks' diet, medication and living conditions (to the extent that such care did not cut into profits – the quintessential challenge of industrial livestock production). Since farmers were paid for the number of live chickens delivered, requirements to make chickens live were effectively written into the contracts. Contracts for FFV similarly spelled out the fertilization and pest control regimes needed for a specified quantity and quality of crops. Notwithstanding this concern with the biological, Watts still treated failures to 'make live' as a problem for grower livelihoods. Watts, that is, analysed contract farming as a form entailing social exploitation, with scant attention to its ecological dynamics (Galt 2014), even though else-where he was highly attentive to the causal connections between the dynamics of capitalist growth and specific environmental outcomes (Peet and Watts 1993; Watts 1987).

In regard to those ecological dynamics, Grossman (1998, 29) writing on contract banana production noted that 'one of the hallmarks of con-tract production is the intensive use of agrochemicals', since, depending on the nature of the contract, it may directly 'stipulate the nature, extent and timing of pesticide use' or, alternatively, indirectly induce pesticide use through quality standards. Hearkening back to Watts' (1987) 'simple reproduction squeeze', others suggested that contract farmers are more prone to more intensive pesticide regimes because as small, under-capitalized producers they farm on marginal land where environ-mental conditions make plants most attractive to pests, giving them little choice but to use pesticides (Clapp 1988; Galt 2014; Morvaridi 1995). Furthering their reliance on pesticides, these farmers do not have a capital cushion for experimentation nor do they have access to the same knowledge and technology that more capitalized and networked farmers do (also Bell 2010; Wells 1996).

Contract farming does not necessarily increase pesticide use, how-ever. In Costa Rica, Galt (2014, 83) found that export contracts required producers to comply with US pesticide residue limits. They also helped producers reduce pesticide use by offering fixed prices for their crops. This, combined with export producers' better access to suitable land for their crops (due to better access to capital), means that in Costa Rica, pesticides are applied most heavily on FFV destined for the domestic market.

These observations on contract farming hold true for California's strawberry industry, where the producers particularly inclined to use pesticides are often those who farm under conditions that reflect and contribute to their economic marginality. I want to illustrate this phenomenon before moving to the biopolitical responsibilities they face.

From Sharecropping to Partnering in California's Strawberry Industry

In her groundbreaking work on the California strawberry industry, Miriam Wells (1984, 1996) shared Watts' concerns with the agrarian question as well as his appreciation for Kautsky's prescience. She questioned the linearity and inevitability in capitalist development trajectories, especially in agriculture. For Wells, a defining moment in the strawberry industry came with the late 1960s resurgence of sharecropping, after it had largely gone away. Growers subdivided their land into 3–5 acre plots and leased them out to croppers (generally of Mexican or Japanese origin), rather than employing workers directly. Their embrace of this 'feudal'-like organization of production made no sense according to theories of development that predicted its inexorable demise. Working through several possible explanations, she concluded that sharecropping re-emerged in response to changes in the labour market, precipitated by, among other things, the expansion of labour-protective laws and the growing presence of the United Farm Workers' union. Sharecropping allowed growers to maintain control over product quality and labour performance while delaying labour payments and reducing overall labour costs, due to croppers' willingness to exploit unpaid family labour.

Strawberry sharecropping in California all but disappeared in the wake of a 1981 settlement of a long-running lawsuit against Driscoll's, the state's most prominent shipper. After the court decided that the company's tight control over production practices made its croppers more akin to wage labourers than independent farmers, the proportion of sharecropped strawberry acreage shrunk from about 50% to 10% (Wells 1996, 270). According to Wells, sharecropping persisted only in forms unlikely to face legal challenges. Among them were the 'partnerships' forged between shippers' intermediaries and growers, in which intermediaries provided financing and market access, but, as Wells saw it, rarely informed growers of the risks involved.

In the early 1990s, Wells (1996) noted the relatively small size of berry farms – typically under 100 acres – which she attributed to the absence of economies of scale as well as high per acre profits. She also observed a wide array of buyers, ranging from large brand-name shippers, some of whom financed and advised their growers, to growers' marketing cooperatives (two major ones at the time), to a number of independent shippers. More than 20 years later, the typical berry farm is much bigger: many growers operate multiple ranches and in multiple regions, and a significant number farm more than 500 acres per year, a huge capital undertaking given the per acre cost of about USD 50,000 (Bolda, Tourte, Klonsky and De Moura 2010). They have acquired this acreage as others have gone out of business due to industry volatility and the decline in per acre profits. Meanwhile five large corporate shippers (one a former cooperative), now dominate the industry, and the remaining cooperative and independent grower-shippers are on the decline.[2]

Most of these shippers both farm their own land and buy from other growers. The important exception is Driscoll's, which produces only on experimental plots. Interview data indicates that the contracts between shippers and growers vary considerably. (Exactly how is hard to say since growers are sworn to secrecy.) On one end of the risk-reward spectrum, 'custom growers' are paid by the acre or through a management fee, regardless of their harvests. On the other end, so-called independent growers sell to the shippers on marketing contracts. Paid by the box at a fluctuating market rate, they face much higher risks but also the possibility to sell at higher prices if, for example, they can harvest early in the season. Offsetting this advantage are buyers' strict quality grades and, in some cases, burdensome fees. Growers selling to Driscoll's and Well-Pict, for example, must pay licence fees for their proprietary varieties, as well as steep sales and cooling commissions (18% in the case of Driscoll's, according to an interviewed grower).

Partner growers (also referred to as associates), like sharecroppers, have neither the security of the custom growers nor the potential for high profitability of the independent growers and yet are showcased as the new face of farming. Driscoll's works with at least two 'affiliated' companies that cultivate such partnerships. They look for particularly entrepreneurial and competent ranch managers, farmworkers and other hired hands and tap them to go into business with the company. Since most of these chosen partners could not otherwise afford to start a business, one of the affiliate companies boasts of its ability to provide such opportunities, joining others in California who attempt to improve the lot of farmworkers by making them farmers. This approach thus reflects a persistent Jeffersonian agrarian ideal (Brown and Getz 2008; Guthman

2004; Schlosser 1995); as Watts (1993) himself noted, contract farming in the United States has been touted as a way to save the family farm.

To establish a partnership the two parties set up a new company, generally a limited liability company (LLC), in which the managing partner puts up 70% of the capital and the grower partner puts up 30%. A grower generally begins with about 20 acres in production, which requires an upfront investment of about USD 150,000, or USD 18,000 per acre (for lease costs, soil preparation, irrigation, planting and so forth). Some growers take second mortgages on their homes; others go to the bank. Either way, they depend on their managing partners to help secure loans for which they would not otherwise be eligible. If and when growers begin to make money, they increase their acreage as well as equity in their respective companies, up to 49%. Unusually successful associates may eventually take majority ownership or even become entirely independent of the affiliates and start working directly with Driscoll's.

Many such growers do not succeed, for several reasons. First, they carry heavy and ever-growing debts. If they lose money, they must borrow to stay afloat and pay off debts to both Driscoll's and their managing partners. If they make money, they borrow anyway to maintain good credit. Second, these resource-poor growers depend on the managing companies to secure not only financing but also land, which they sublease from the companies. In the tight land markets of California's strawberry regions, this might seem like a valuable service, but it also a less than transparent one. Some partner growers suspect that the managing partners allocate the best land to their favoured growers, and may also engage in arbitrage. Third, the partner growers must pay Driscoll's for almost all inputs, including licence fees for its proprietary strawberry varieties. While these varieties often sell for higher prices, presumably due to their quality, the fees cut into any extra earnings. Partner growers must also rent equipment from the companies, buy Driscoll's marketing materials (such as labelled boxes and baskets) and pay them management fees as well as sales commissions. Crucially, these expenses all return to Driscoll's and its affiliate companies before any residual revenue is made and divided. Fourth, these partner growers must comply with the same stringent grading standards as Driscoll's independent growers, but often under much less favourable conditions, due to their economic marginality. Many of them farm in Oxnard, for example, on land with insufficient or poor quality water. Because water-stressed plants are more susceptible to spider mites and soil pathogens, these growers must spend more money on pesticides and fertilizers than they otherwise would.

In short, compared to California's other strawberry growers, these partner growers typically contend with higher risks and tighter margins, as well as more rent-seeking on the part of their managing partners and Driscoll's. I contend that these rent-seeking opportunities, rather than the avoidance of labour laws, are one of the primary, albeit under-recognized, advantages of these arrangements, in addition to expanding Driscoll's productive base (cf. Schlosser 1995; Wells 1996 and many critics from within the industry). They certainly do not benefit the growers, who seem to go out of business quite regularly, often saddled with a great deal of debt. Thus far they have been easily replaced by others who believe they will fare better. As put by one outside observer, 'Driscoll goes through growers like my kid goes through socks.'

In my view, the model is unworkable not only for economic reasons, but also because these contracts oblige growers to manage life and death under competing, even impossible imperatives. These are contract farming's biopolitical dimensions, to which I now turn.

Toward a Biopolitics of Contract Farming

Remarkably, most of the literature on contract farming, including Watts' significant contributions, has highlighted its effects on grower livelihoods, including that which has focused on the biological risks of agriculture that pesticides presumably mitigate. Taking issue, Galt (2014) noted that dependence on pesticides not only undermines future conditions of production, as pests become resistant to their use and soil quality erodes; it also puts farmers, workers and the surrounding community in harm's way. Still, in discussing the human cost of pesticide use, Galt treated the problem as somewhat epiphenomenal. A 'sad irony of pesticide use,' Galt (2014, 210) wrote, 'is the unintentional displacement of disease from one organism (the crop plant) to another (humans)'. In contrast, a biopolitical analytic recognizes that sacrificing some life is intrinsic to processes which make life live.

Foucault introduced the term biopower to note a shift in the role of government in modern states towards the assurance of 'life itself' distinct from, although not unrelated to, profit-making and other economic activities. Food production is therefore inherently a biopolitical undertaking: it is about making plant and animal life live in order to make human life live. But as Foucault and others have emphasized, to make life live requires identifying which life is important and then removing the threats to that life. Foucault (1985, 1997, 2007), of course, was writing about human life. He differentiated the 'population' from the 'people', by suggesting that the former is constitutive of those whose lives count while the others are put to die or let die, depending on whether they are threats or simply not necessary. Recent scholarship has extended

Foucault's notion of the population beyond human life, noting how human management of plants, animals and other organisms – whether in the form of saving endangered species or making cheese – never aims for all life to flourish. Rather, it often entails the neglect or eradication of any life detrimental to valued life (Biermann and Mansfield 2014; Bobrow-Strain 2008; Collard 2012; Lorimer and Driessen 2013; Paxson 2008). Pesticides are in that way classic 'technologies of security' – those that minimize or eradicate threats on behalf of the population (Dillon and Reid 2009; Foucault 1997; Foucault 2007). Clearly, though, pesticide use subjects not only crops and their pests to biopolitical sorting, but also farm workers, neighbours, farmers and consumers.

In the context of California's strawberry industry, contract growers face the problem of *competing* biopolitical exigencies. On the one hand, contracts themselves make the life of the strawberry plant paramount, and demand that growers produce the strawberry both as a commodity and as 'healthy' food for the consuming population. On the other hand pesticide regulations, technologies of security in their own right, serve to protect the health of the neighbouring population.

The use of soil fumigants puts these competing exigencies into stark relief. Strawberry growers rely heavily on soil fumigants not only to control weeds and nematodes but also to disinfect fields of soil pathogens that kill plants by preventing their nutrient uptake. But due to their high toxicity, fumigants have seen much tighter regulatory scrutiny of late and two, methyl bromide and methyl iodide, are no longer available. Like other technologies of security, these new restrictions inherently privilege some lives for others. To be sure, the question of which human lives shall be protected and how has been at the heart of much of the regulatory controversy over fumigant use in California. For example, fumigant mitigation measures protect farms' neighbours much more effectively than their workers (Guthman and Brown 2016), while pesticide regulations more broadly have protected consumers more than farming communities, in part by encouraging chemical formulations that leave no toxic residues (Harrison 2011; Wright 2005). Yet emerging fumigant reduction regulations and plans hint at biopolitical recognition for workers as well as neighbours. For instance, in 2013 California's Department of Pesticide Regulation released a report about the need to curtail and eventually phase out fumigants to protect the health of farmworkers, bystanders and nearby communities (Department of Pesticide Regulation 2013). In 2015, EPA released new worker safety rules for pesticide applications. These new regulatory developments complicate the work of contract growers, in that they do not explicitly forbid pesticide use but rather require growers to employ increasingly burdensome mitigation measures while still delivering 'quality' berries at a low cost.

But it is very difficult to reduce, much less forego fumigation when plants are stressed. At the same time, growers must continue to earn enough to keep themselves in business and their workers employed, or else they effectively let humans die in other ways, from the life-shortening consequences of destitution. Even when employed, many farm workers contend with sub-par housing and undernourishment (Carney 2014).

In short, the competing demands of contracts and regulations require growers to make life and death decisions under intense economic pressures. Contracting out production allows companies such as Driscoll's, meanwhile, to avoid the costs of regulatory compliance, including those resulting from yield losses. It also makes it easier for them to distance themselves from the moral baggage of fumigant use, which has come under increasing public criticism. Driscoll's, in fact, seeks to be a leader in reducing the industry's dependence on fumigants, and has invested significant resources in researching alternatives. While its contracts do not forbid fumigant use, the company's public stance sends a message to out-growers to consider farming without them – though it provides no direct financial assistance to help with this transition. In effect, the contracts devolve the impossible responsibility for securing multiple forms of plant and human life to the farmer while allowing Driscoll's to take the high road. This, I argue, is another key, under-recognized advantage of these contract arrangements for Driscoll's and the managing partners.

* *

To conclude, all growers must kill life to make life – that is the nature of farming which necessarily contends with the biological, as Watts emphasized in his work on contract farming. In classic agrarian political economy, an important benefit of a production contract for managing firms is that it leaves these biologically risky aspects of production to the most economically marginal farmers whose failure to succeed in those circumstances becomes a risk to their livelihoods. Human life remains external to the analysis, even in most of the scholarship addressing pesticide use in such systems. But human life is also at risk in these pesticide-dependent production systems, and depending on which pesticides are used and how they are applied, different human lives are at risk. A biopolitical analysis thus offers a different vantage point. It recognizes that not all lives can be made to live, and points to the sorting and letting die needed to make favoured life live. This truth could not be starker in the case of food production, which involves all sorts of killing in the service of the population. When supermarket standards, state regulations and popular sentiment dictate different ideas of whose and what lives should be protected, and therefore whose and what lives might be sacrificed, those who must follow those dictates are put in a near impossible position. At the same time, shippers and other intermediaries, while not entirely absolved of biopolitical responsibility, can potentially benefit – at least reputationally – from their commitments to move beyond pesticides.

In the final analysis, though, the shippers and intermediaries are not only offloading responsibility. Put starkly, the managing partners in these sharecropping arrangements may share some proceeds, but they do not share pesticide exposures. And so, in their efforts to meet all exigencies prescribed by the contract, partner growers do more than compromise their livelihoods and likely the lives of those around them; in creating a healthy berry they take on life risk itself.

**

Thank you, Michael, for both your generosity in teaching and prodigiousness in research that has provided so much fodder to think with.

Notes

1   It is not clear that this claim has held up to time, especially in light of the huge wave of land grabs that have taken place since the early 2000s.
2   At the same time direct marketing has seen a substantial uptick, driven by the organic and farmers' market booms.

References

Bell, M. M. 2010. *Farming for us All: Practical agriculture and the cultivation of sustainability*. State College, Pennsylvania State Press.
Biermann, C. and Mansfield, B. 2014. Biodiversity, purity, and death: conservation biology as biopolitics, *Environment and Planning D: Society and Space*, vol. 32, no. 2, pp. 257–273.
Bobrow-Strain, A. 2008. White bread bio-politics: purity, health, and the triumph of industrial baking, *Cultural Geographies*, vol. 15, no. 1, pp. 19–40.
Bolda, M., Tourte, L., Klonsky, K. and De Moura, R. L. 2010. *Sample Costs to Produce Strawberries, Central Coast Region*. UC Cooperative Extension.
Boyd, W. and Watts, M. J. 1997. Agro-industrial just-in-time: the chicken industry and postwar american capitalism, in D. Goodman and M. J. Watts (eds), *Globalising Food: Agrarian questions and global restructuring*. London, Routledge.
Brown, S. and Getz, C. 2008. Towards domestic fair trade? Farm labor, food localism, and the 'family scale' farm, *Geojournal*, vol. 73, 11–22.
Carney, M. A. 2014. The biopolitics of 'food insecurity': towards a critical political ecology of the body in studies of women's transnational migration, *Journal of Political Ecology*, vol. 21, 1–18.

Clapp, R. A. J. 1988. Representing reciprocity, reproducing domination: ideology and the labour process in Latin American contract farming, *The Journal of Peasant Studies*, vol. 16, no. 1, 5–39.

Collard, R.-C. 2012. Cougar–human entanglements and the biopolitical un/making of safe space, *Environment and Planning-Part D*, vol. 30, no. 1, pp. 23–42.

Department of Pesticide Regulation. 2013. *Nonfumigant Strawberry Production Working Group Action Plan*. California Department of Pesticide Regulation.

Dillon, M. and Reid, J. 2009. *The Liberal Way of War: Killing to make live*. London, Routledge.

Foucault, M. 1985. *History of Stroduction*. New York, Vintage.

Foucault, M. 1997. *Society must be defended: Lectures at the Collège de France, 1975–1976*. New York, Picador.

Foucault, M. 2007. *Security, territory, population: Lectures at the Collège de France 1977–1978*. New York, Picador.

Galt, R. E. 2014. *Food Systems in an Unequal World: Pesticides, vegetables, and agrarian capitalism in Costa Rica*. Tucson, AZ, University of Arizona Press.

Goodman, D., Sorj, B. and Wilkinson, J. 1987. *From Farming to Biotechnology*. Oxford, UK, Basil Blackwell.

Goodman, D. and Watts, M. J. (eds). 1997. *Globalising Food: Agrarian questions and global restructuring*. London and New York, Routledge.

Grossman, L. S. 1998. *Political Ecology of Bananas*. Chapel Hill: University of North Carolina.

Guthman, J. 2004. *Agrarian Dreams: The paradox of organic farming in California*. Berkeley, University of California Press.

Guthman, J. and Brown, S. 2016. Whose life counts: biopolitics and the 'bright line' of chloropicrin mitigation in California's strawberry industry, *Science, Technology & Human Values*, vol. 41, no. 3, pp. 461–492.

Harrison, J. L. 2011. *Pesticide Drift and the Pursuit of Environmental Justice*. Cambridge, MA, MIT Press.

Kautsky, K. 1988. *The Agrarian Question*. London, Zwan Press.

Lorimer, J. and Driessen, C. 2013. Bovine biopolitics and the promise of monsters in the rewilding of heck cattle, *Geoforum*, vol. 48, pp. 249–259.

Morvaridi, B. 1995. Contract farming and environmental risk: the case of Cyprus, *The Journal of Peasant Studies*, vol. 23, no. 1, pp. 30–45.

Paxson, H. 2008. Post-pasteurian cultures: the microbiopolitics of raw-milk cheese in the United States, *Cultural Anthropology*, vol. 23, no. 1, pp. 15–47.

Peet, R. and Watts, M. J. 1993. Development theory and environment in an age of market triumphalism, *Economic Geography*, vol. 69, no. 3, pp. 227–253.

Schlosser, E. 1995. In the strawberry fields, *Atlantic Monthly*, vol. 276, no. 5, November, pp. 80–108.

Watts, M. J. 1987. Drought, environment and food security, in M. Glantz (ed.), *Drought and Hunger in Africa*. Cambridge, UK, Cambridge University Press.

Watts, M. J. 1993. Life under contract: contract farming, agrarian restructuring, and flexible accumulation, in M. J. Watts and P. Little (eds), *Living under Contract*. Madison, University of Wisconsin Press.

Wells, M. 1984. The resurgence of sharecropping, *American Journal of Sociology*, vol. 90, 1–29.

Wells, M. 1996. *Strawberry Fields: Politics, class, and work in California agriculture*. Ithaca, NY, Cornell University Press.

Wright, A. 2005. *The Death of Ramón González: The modern agricultural dilemma*. Austin, TX, University of Texas Press.

7

Commoditization, Primitive Accumulation and the Spaces of Biodiversity Conservation

Roderick P. Neumann

Political ecologists are producing an illuminating body of research on the material and symbolic roles of modern nature conservation. Within this literature, I draw specific attention to studies investigating a range of inter-related phenomena labelled as 'green grabbing', 'neoliberal conservation', and 'conservation displacement' (e.g. Brockington and Igoe 2006; Agrawal and Redford 2009; Fairhead, Leach & Scoones 2012; Benjaminsen and Bryceson 2012; Büscher, Sullivan, Neves, Igoe, & Brockington 2012; Ojeda 2012). Collectively these works strive to explain how conservation is shaped by or contributes to capitalist development under neoliberalism and to understand the consequences for rural lives and livelihoods. In this chapter I suggest that such inquiries can profit from a deeper historical perspective and from revisiting political ecology's agrarian studies roots, specifically Michael Watts' earliest, path-breaking works (Watts 1983a, 1983b, 1987). Using archival sources, published historical accounts and secondary literature I focus on southeastern Tanzania from roughly 1850 to 1950.[1] During those decades the region's peasant producers were key participants in a series of booms in globally traded extractive commodi-ties. I argue that the successive booms in elephant ivory, rubber and beeswax produced the political-ecological conditions that made possible the displacement of approximately 10 000 African peasants and the estab-lishment of Africa's largest wilderness area, the Selous Game Reserve (GR).

Other Geographies: The Influences Of Michael Watts, First Edition. Edited by Sharad Chari, Susanne Freidberg, Vinay Gidwani, Jesse Ribot and Wendy Wolford.
© 2017 John Wiley & Sons Ltd. Published 2017 by John Wiley & Sons Ltd.

The Argument

A new sub-genre within political ecology has emerged following 'a veritable explosion of scholarship examining the neoliberalization of ... conservation' (Fairhead et al. 2012, 240). The self-declared 'epistemic community' of neoliberal conservation scholars investigates how nature conservation under neoliberalism has increasingly become an important source of capital accumulation (Brockington and Duffy 2010, 479). An exemplary proposition guiding the investigations posits: 'the international biodiversity conservation agenda has created new symbolic and material spaces for global capital expansion' (Corson 2010, 579). In theorizing the increasing opportunities for private profit associated with conservation territories, some have turned to Marx's (1976) concept of primitive accumulation and Harvey's (2003, 2005) reformulation, 'accumulation by dispossession' (e.g. Kelly 2011; Corson 2010; Ojeda 2012; Neves and Igoe 2012; Benjaminsen and Bryceson 2012). Based on Marx's (1976, 875) statement that primitive accumulation 'is nothing else than the historical process of divorcing the producer from the means of production', the key argument here is that primitive accumulation is characterized by the enclosure of a commons and thus the enclosure of conservation territories can be viewed as a source of accumulation. Some authors have observed that, unlike the historic acts of enclosure in England that Marx analysed, conservation enclosures take land out of agricultural production and do not involve the privatization of land and resources, but rather the conversion from individual or common ownership to state ownership (e.g. Kelly 2011; Benjaminsen and Bryceson 2012). The path out of this paradox is paved with a series of ill-fitting analogies. Specifically, the accumulation of *capital* prior to capitalism that Marx theorized is equated with new opportunities for rent seeking by state actors, increased revenue for non-profit organizations, the production and sale of new visual media products, and new investment sites for domestic and international tourism industries (e.g. Corson 2011, Kelly 2011; Benjaminsen and Bryceson 2012).

There seems no doubt that conservation enclosures can, either directly or indirectly, provide new sources of profit or revenue for a range of social actors. The question, rather, is whether contemporary conservation enclosures are a form of primitive accumulation and whether neoliberal conservation constitutes some new mode of production (see Garland 2008; Brockington and Scholfield 2010). It can be productive to both step back from the present neoliberal moment toward a *longue durée* approach to conservation territory formation and simultaneously revisit political ecology's foundational texts. The theoretical concerns common in much of the neoliberal conservation literature – primitive accumulation, enclosures and proletarianization – were the concerns of the geographers and anthropologists whose agrarian transition research helped establish political ecology. Foremost among these foundational texts are Watts' publications resulting from his historical and ethnographic research in Northern Nigeria.

Watts' critical examination of the relationship between drought and famine in Northern Nigeria rests on his historical analysis of the region's incorporation into the world capitalist economy under European colonialism. In particular, he targets the historical 'process of commoditization and the social context of the development of markets among differentiated users' (Watts 1987, 172). Colonialism, he argues, 'progressively, but in highly heterogeneous ways, transformed peasant households into petty commodity producers, the central characteristic of which is the circulation of commodities in both directions' (Watts 1983a, 20). Through this focus on commoditization he highlights differentiation, heterogeneity and unevenness as qualities inherent in the processes of capitalist incorporation. 'In Northern Nigeria, the process of primitive accumulation remained only differentially completed' (Watts 1983a, 22). This latter observation illustrates his conceptualization of primitive accumulation as an historically extended, incomplete process rather than as a singular event, such as an act of enclosure. Indeed, Northern Nigerian did not experience enclosure, forced labour or the transformation of peasant production. 'Colonial integration did nonetheless initiate the process of primitive accumulation, which projected Nigeria into a global division of labor ...' (1983a, 22). This conceptualization reflects Marx's understanding of primitive accumulation 'as an accumulation which is not the result of a capitalist mode of production but its point of departure' (1976, 874).

Equally important to his historical analysis is his ethnographic approach to understanding farmers' responses to conditions of drought as they relate to the emergence of famine. By concentrating on rural social relations, Watts reasoned, 'one can begin to understand the connections between material circumstances and ecological conditions' (1987, 190). Building on this work, I analyse the origins of the Selous GR as a means to reassess the relevance of primitive accumulation to conservation practice and, more broadly, to place conservation territories within an historically unfolding process of uneven capitalist development. My positions are as follows. Contra the idea that contemporary conservation is opening new opportunities for capitalist accumulation, I seek to understand historically how uneven capitalist development produced new material and symbolic spaces of opportunity for international conservation interests. Second, rather than focusing on incidents of enclosure only, I view primitive accumulation as a decades-long process of deepening commoditization that may or may not include enclosures. Third, by refocusing on land managers, I stress the agency of peasant producers in helping to shape the political-ecological conditions of their own displacements.

The Location

The study area lies between 7.5 degrees and 10 degrees south latitude and extends inland from the Indian Ocean's western coastline to approximately 37 degrees east longitude (Figure 7.1). It encompasses roughly

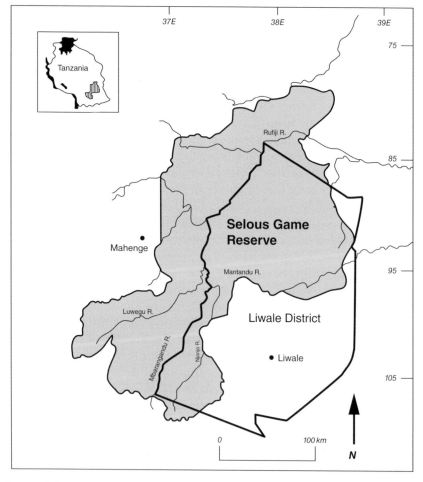

Figure 7.1   Location and boundaries of present-day Liwale District and the Selous Game Reserve in Tanzania.

200 000 square kilometres of the southeastern corner of Tanzania. The terrain is predominantly rolling, forested hills – called *miombo* woodland after the Kiswahili name for the dominant tree species (*Brachystegia* spp.) – but also includes open grasslands, riparian forests and wetlands. Most of the land ranges from 300 to 700 metres elevation and is frequently bisected by numerous seasonal and perennial streams and rivers. Many of the larger valleys have deep alluvial soils, while the uplands tend to have shallow soils of lower fertility. Rainfall averages 600 to 800 millimetres annually, punctuated by a dry season from May to October. Many wildlife species occupy the area, notably a large population of African elephant (*Loxodontha africana*).

Caravan routes that connected the interior to the coastal ports of Kilwa and Zanzibar had for centuries linked the Kilwa hinterland – as this region is historically known – to transoceanic trade (Iliffe 1979). In the nineteenth century, 'legitimate' trade brought East Africa into closer contact with European traders and explorers. At that time the area was mostly occupied by the Ngindo people, a stateless, primarily linguistically defined tribal assemblage.[2] The current scholarly consensus categorizes early nineteenth-century Ngindo politics as clan or chieftaincy based, and their economy as primarily based in permanent and semi-permanent valley cultivation supplemented by hunting and gathering in surrounding forest and bush lands (Kjekshus 1977; Iliffe 1979; Koponen 1988; Giblin and Monson 2010).

Following the 'outburst of imperialism' in Europe in the 1870s, this portion of East Africa experienced greater world-economic integration culminating in formal European colonization, first by Germany (1885–1916) and then by Great Britain (1916–1961) (Great Britain Foreign Office 1920a, 28). German colonizers introduced hunting and game laws in 1896, which they designed to monopolize the ivory trade and reserve hunting for Europeans at the expense of Africans' access (Sunseri 2010). They divided the territory into administrative districts, with Ungindo comprising most of the Kilwa District, and appointed Swahili and African *akidas* (village headmen) to serve as local administrative intermediaries. Many *akidas* lacked political legitimacy among their subjects but were responsible for enforcing unpopular government regulations, mobilizing collective labour for government projects, and collecting a hut tax, which the Germans introduced in 1897.

British colonizers built largely upon German government structures, reproducing an administrative hierarchy of provinces and districts. The Germans had established two small game reserves for European hunters in the north of Ungindo in 1905, which the British later incorporated into the Selous GR. In 1925 the British introduced from Nigeria a system of indirect rule, which established African 'native authorities' comprised of chiefs and *jumbes* (village headmen). They placed Ungindo under the Ngindo Native Authority, whose boundaries overlapped with those of the Liwale District, which fell within the Southern Province.[3] What follows is a conservation territory origin story featuring Wangindo participation in a series of commodity booms and the unintended political-ecological consequences of that participation.

Ivory

Elephant ivory was a key commodity driving the deeper integration of East Africa into the global economy during the decades preceding formal colonization. The rapid growth and international spread of bourgeoisie culture in the nineteenth century spurred global demand for ivory piano keys, combs, billiard balls and myriad tchotchkes (MacKenzie 1988). Demand generated a tremendous expansion in the regional caravan trade, making East Africa the single largest source of ivory in the world (Beachey 1967; Hakansson 2004). Ivory flowed through the caravan tributaries in the interior to East Africa's Indian Ocean trading ports of Kilwa and Zanzibar, where American and European firms purchased it and transported it across the seas for processing. By the latter part of the nineteenth century, the ivory boom had transformed social, economic, political and ecological conditions in much of East Africa. Historians have calculated that the 10 000 porters moving annually along one central Tanzania caravan route needed a minimum of 400 metric tons of increased crop production to sustain them (Hakansson, Widgren and Börjeson 2008). Expanding caravan trade made the consumption of luxury commodities – cloth, beads and metal implements of all sorts – commonplace among agriculturalists (Rempel 1998; Hakansson et al. 2008). Ivory was a ready source of wealth for African chieftaincies and states, which were then able to acquire guns and slaves and thus consolidate and expand control (Rempel 1998).

In Ungindo, African peasants' participation in the ivory boom altered the local political ecology. Wangindo agriculturalists were ideally situated to expand crop production as caravan traffic increased along an ancient route that bisected their territory (Koponen 1988; Larson 2010). A first-person account from 1860 reported that Ungindo was 'extremely well settled and cultivated in an excellent manner' and that residents offered 'goats, chicken, peas, beans, millet, sweet potatoes, flour, sugar cane, mangoes, and pistachio nuts for sale' (Decken, quoted in Kjekshus 1977, 73). Wangindo men were skilled hunters and masters at concocting vegetal poisons for arrows and spears, which they also produced for trade. In particular, skill in elephant hunting may have been central to the formation of political identity in the Kilwa hinterland, including Ungindo. Local big men emerged among the most skilled Wangindo elephant hunters who were able to trade ivory to accumulate and redistribute imported commodities, especially cloth (Sunseri 2010). The consequent decline of elephants in the landscape had significant ecological implications, most notably a reduction in grasslands and an increase in bush and woodlands. Ecologists classify the African elephant as a 'keystone species', meaning that it plays a central role in habitat transformation for numerous other plant and animal species. By the 1890s, hunting had reduced and scattered the elephant herds and very few, if any, elephants remained in Ungindo (Matzke 1977; Rodgers 1976).

Rubber

By 1890 the trade in ivory had peaked, but another extractive commodity, wild rubber, had emerged to take its place. Rubber, extracted from two indigenous woodland plants, one a vine (*Landolphia stolzii*) and the other a shrub (*L. tondeensis*), had long been exported from Kilwa and Zanzibar (Great Britain Foreign Office 1920b, 66). Then in the late nineteenth century global demand for East African rubber became unquenchable for several decades running. By 1876 it was Kilwa's chief export and soon surpassed the value of ivory exports (Iliffe 1979; Larson 2010). At the height of the Tanzania rubber boom, only Brazil exceeded Zanzibar as a global source of rubber (Iliffe 1979).

Ungindo was one of two regional sources for Tanzania's exported wild rubber and Ngindo collectors became important suppliers. Collectors preferred to carry the load directly to wholesale exporters in Kilwa where they could find the most favorable exchange rates.[4] Wangindo commonly traded rubber for cloth, which they in turn traded for slaves in order to expand cultivation (Iliffe 1979). A self-reinforcing cycle of expansion was thereby established in which rubber profits were used to purchase labour (in the form of slaves), which was used to expand grain production to supply an export market, which led to further increases in caravan traffic and further demand for grain. Consequently, grain exports, measured in caravan loads from Kilwa, increased rapidly at the turn of the twentieth century. From 1897 to 1903, annual loads jumped from 5050 to 11 334, with Ungindo accounting for about one-quarter of the latter total (Iliffe 1979).

As wealth generated by the boom flowed into Ungindo, Wangindo men and women consumed imported cloth as well Islamic Swahili clothing fashions from the coast and European fashions from abroad (Iliffe 1979; Larson 2010). Then, as now, household consumption of luxury goods was often fuelled by credit and accompanied by rising personal debt. Those who chose not to make the trip to Kilwa dealt with itinerant Arab and Indian traders who delivered goods on credit, the debt to be paid in interest with rubber. The government, meanwhile, had introduced a variety of regulations (e.g. mandatory labour on European plantations for tax arrears) that produced internal political tensions within Ungindo, increased animosity toward the Germans, and spurred peasants to devise tax and labour avoidance tactics. Tensions eventually built to the point of rupture.

The Maji Maji Rebellion of 1905–1907, the largest anti-colonial uprising in Tanzania, abruptly ended rising prosperity in Ungindo. The fact that the largest rebellion originated in the one region of the colony that was prospering suggests the contradictions of capitalist social relations had produced a crisis that 'originated in peasant grievances' (Iliffe 1967, 495). Wangindo participants and observers mentioned several grievances, including game laws, hut taxes, corporal punishment and forced labour rules.[5] It is noteworthy that 'the moving spirit of the Rebellion in the central [Ungindo] area was an [Ngindo] elephant hunter named Abdallah Mapanda'.[6] After German game laws curtailed Ngindo hunters' access to ivory, poaching and smuggling became widespread. Significantly, African elephant hunters were key rebellion leaders in Liwale and elsewhere (Sunseri 2010).

The political-ecological consequences of the extractive commodity booms and the subsequent rebellion and its aftermath were extensive. The government hanged rebel leaders and rewarded loyalists, thus reconfiguring political power within Ungindo. The demographic effects were dramatic because the Germans employed scorched earth tactics, thereby inducing famine (Iliffe 1967, 1979; Redmond 1975). Estimates put the total population loss at 195 000 to 300 000 or about one-third of the region's population (Great Britain Foreign Office 1920b; Iliffe 1979, 200). African peasants never reoccupied many areas that were densely settled and extensively cultivated in the late nineteenth century.[7] Woodland and bush expanded and wildlife began to populate areas abandoned by farmers and their livestock (Iliffe 1979; Kjekshus 1977). A decade or so after the rebellion, elephants were seen in the region for the first time in living memory (Rodgers 1976).

Beeswax

Even without Maji Maji's crushing impacts, the rubber boom in Ungindo was doomed. In 1913, prices for East African rubber collapsed when Southeast Asian sources of higher quality Brazilian rubber (*Hevea brisiliensis*) entered the global market. Once again, another extractive commodity boom was expanding even as one faded. At the beginning of the twentieth century global beeswax prices were on the rise as demand from new industrial processes was added to expanding demand for established uses. According to colonial agronomists, Ungindo included some of the most productive honey and beeswax country in East Africa, for they judged *miombo* woodland to be 'composed almost entirely of nectar producing species'.[8] In addition, the region's geomorphology of rolling hills bisected by numerous watercourses provided the proximity to water that is essential for bees.

Some Wangindo men historically specialized in beekeeping. Called 'bee *fundis*' (Kiswahili for skilled craftsperson) by Europeans, their traditional focus had been on honey. In the decades immediately preceding German occupation, however, beeswax emerged as an East African export commodity and *fundis* shifted their focus. Individual beekeepers might manage hundreds of hives, as many as 1000 according to one report.[9] Exports from German East Africa exploded from 1.6 tons in 1889 to 100 tons in 1897 as exporters scurried to meet rising global demand (Tuck 2009). At the height of the trade, only the United States supplied more of the world's beeswax.[10] In 1933, the British exported 16.6 tons out of Kilwa, with 75% of the total coming from Ungindo.[11] In a 1942 survey of the Ngindo Native Authorities, comments for nearly every settlement included assessments such as: 'In good years ... a considerable amount of wax' and '... in most years beeswax is plentiful'.[12]

To colonial officials' frustration, however, beeswax extraction depended on peasant mobility, making administration and taxation more challenging. As with rubber collectors, many Wangindo beekeepers acquired cash through beeswax sales and thus rejected wage labour.[13] And as with rubber collectors, bee *fundis* preferred to market their wax directly to Kilwa exporters. Government authorities noted: 'After the harvest a general exodus of the able-bodied safari to the coast', which beekeepers laden with wax joined.[14] Authorities' frustration lay precisely with the vastness of the beekeepers' domain and the coastal safaris: 'unfettered movements of the native and the uncertainty of his return from the Coast render [tax] evasion easy.'[15] While beeswax continued to be an important Tanzanian export into the 1950s and beyond, Ungindo's contribution declined due to colonial officials' solution to the problem of what they called Wangindo's 'wanderlust.'[16]

Displacement

Upon taking control of German East Africa, the British government became aware that elephant populations had rebounded enough to threaten the viability of peasant agriculture. In 1921 an 'organised campaign was started against these raiders' by the government (Great Britain Colonial Office 1923, 7). Authorities temporarily relaxed the game laws, gave away elephant licences to European hunters and distributed powder and caps to African owners of muzzle-loading guns. By the 1920s game rangers were killing 800 elephants a year across Tanzania, but herds continued to expand. In response to food shortages and excessive crop losses the government inaugurated in 1933 an elephant control scheme in the Southern Province.[17] The scheme involved driving elephants westward deeper into the Kilwa hinterland and eradicating them nearer the coast.[18] In the first year of the scheme, game rangers killed 1304 elephants in that province alone, mostly in Liwale District.[19]

These efforts to protect and expand African peasant production in Liwale were riddled with contradictions. While Europeans barred Africans from elephant hunting, they were alone unable to check the growth of elephant populations even as crop destruction spread.[20] Moreover, the government was aware of Ungindo's past prosperity and its recent status as a grain-exporting region, yet protection for much of the region's agriculture was refused. The scheme initiated in 1933 required receiving areas for elephants driven westward and small game reserves were created for that purpose. No government crop protection would be provided there, although they were known to include sites of historically productive agriculture. Colonial administrators recognized that river valleys in the new game reserve constituted 'probably the most well-watered and fertile in the District.'[21] However, the 'pressure of elephant,' one official noted, 'is already very great and the natives are finding it difficult to maintain their cultivations.'[22] In essence, a scheme ostensibly intended to rationalize peasant agriculture was making cultivation impossible in some of Ungindo's most productive valleys. This scheme marks the beginning of a modern process of displacement and the separation of producers from the means of production in Ungindo.

By the 1940s, the central authorities viewed the Liwale District and its residents as backward and ungovernable. Tanzania's administrative secretary concluded that the only viable solution was 'resettlement of the bulk of the population'. Once the area was depopulated, it 'should be declared a game reserve'. He cited various problems of administration; difficulties in tax collection, a weak Ngindo Native Authority, travel problems and an inhospitable environment, described as 'infested with tsetse from end to end and teem[ing] with game'.[23] In May 1944, administrators agreed that the evacuated areas would be added to a greatly expanded Selous GR, making it 'a reasonably compact block of about 15,000 square miles'.[24] Many dispossessed peasants turned to migrant labour, a phenomenon that the colonial labour officer claimed had 'scarcely started before 1947' and which he attributed chiefly to the effects of the evacuation.[25] Around the turn of the twentieth century, Ungindo was importing labour in the form of slaves and exporting surplus grain. Fifty years later it was exporting its labour and feeding its surplus agricultural production to wildlife.[26]

Conclusion

The prevailing focus on the new in neoliberal conservation can seem to obviate the need for a *longue durée* approach. Perhaps as a consequence, one also notes a slippage in language that apparently treats conservation as distinct from or even commensurate with the capitalist mode of production, as in 'partnerships between conservation and capitalism' (Igoe, Neves and Brockington 2010, 487). Most provocatively, some have gone so far as to claim 'conservation as a mode of production' (Garland 2008, 61; Brockington and Scholfield 2010). Among other problems, such language risks setting up a conservation–capitalism dichotomy that reproduces the nature–society dichotomy that early political ecology was intent on demolishing (e.g. Watts 1983b). To be sure, some of the same neoliberal conservation authors recognize that conservation was never 'a domain separate and set apart from capitalism' and that neoliberalism is 'but the latest in a long and healthy relationship between capitalism and conservation' (Brockington and Scholfield 2010: 552; Brockington and Duffy 2010, 470). It is not, however, merely a matter of conservation and capitalism sharing a history. Rather, modern conservation, as the case of the Selous GR suggests, is an outcome of the spread and intensification of commoditization over time and the geographic unevenness of capitalist development. As with modern drought and famine, modern conservation 'can only be understood in historical fashion' (Watts 1987, 257). Stated differently, conservation must be understood through an historical analysis of the production of nature and space within the capitalist mode of production.

I used the case of the Selous GR to argue that the spaces of conservation do not necessarily generate the primitive accumulation of capital. Rather, in this case, the process of primitive accumulation – and uneven capitalist development more generally – produced the spaces of conservation. Historical analysis can thus clarify how particular spaces became conservation worthy in the eyes of scientists, resource managers and conservation activists. In colonial Tanzania, African peasants' active participation in the commoditization process helped establish the political-ecological conditions for their eventual dispossession in the name of wilderness preservation. Furthermore, in Ungindo the process of primitive accumulation, as Watts observed in Northern Nigeria, initially 'transformed peasant households into petty commodity producers', rather than landless wage labourers (Watts 1983a, 20). Population displacements and the enclosure of the Selous GR were the unintended outcomes of primitive accumulation, not its initiation.

Notes

1   The research was supported by the National Science Foundation, Geography and Regional Science Program, SBR-96 17798. I use the current country title of Tanzania, which was first German East Africa and then Tanganyika.

2   As Iliffe (1979) and others have noted, tribal identities in East Africa were not discrete entities occupying well-defined territories. The British administrators tended to think of Ngindo as a 'trade name,' meaning traders used it to identify the linguistically similar people occupying the Kilwa hinterland. I use the Kiswahili forms 'Wangindo' to refer to the plural and 'Ungindo' to refer to their territory.

3   Nachingwea District Book, sheet no. 8, Tanzania National Archives; hereafter TNA.

4   Nachingwea District Book, TNA.

5   From testimony gathered in 1940 and oral histories collected in 1968 by the University of Dar es Salaam; respectively, 'The Maji-Maji Rebellion in Liwale District,' Nachingwea District Book, TNA and *Maji Maji Research Project*, TNA.

6   The Maji-Maji Rebellion in Liwale District,' p. 8, Nachingwea District Book, TNA

7   British colonial officials later noted that the area 'was much more thickly populated than it is now...[and that earlier travelers] described as 'nothing but houses just like Dar es Salaam."' 'The Maji-Maji Rebellion in Liwale District,' p. 9, Nachingwea District Book, TNA. A survey of one river valley notes that it has 'remained desolate' since German forces surrounded and destroyed a rebellious community there. 'Lukuliro River-Valley Survey Report,' 23 November, 1942 by A.F. Litchfield. Acc. No. 16, 19/70, TNA.

8   'Synoptic Review of Agriculture: Beeswax and Beekeeping' Acc. No. 16, 26/2 TNA.

9   'Beeswax and Beekeeping' Acc. No. 16, 26/2 TNA.

10   'Beeswax and Beekeeping' Acc. No. 16, 26/2 TNA.

11   D.C. Kilwa to P.C. Lindi, 5 Feb.1934, Acc. No. 16, 26/2 TNA.

12   Nachingwea District Book, 'The Native Authorities of Liwale District,' c. 1942, TNA.

13   'Memorandum on the Concentration of Population in the Liwale District,' September, 1943. Acc. No. 16, 11/204, TNA.

14   Nachingwea District Book: Hut and Poll Tax, TNA. Provincial Commissioner (PC) Eastern to PC Southern Province, 4 Sept. 1935 Acc. No. 16, 26/2, TNA.

15 'Memorandum on the Concentration of Population,' September, 1943. Acc. No. 16, 11/204, TNA.
16 'Memorandum on the Concentration of Population,' September, 1943. Acc. No. 16, 11/204, TNA.
17 Minute, illegible, 4 Aug., 1933. TNA 21726.
18 Game Ranger to Acting Game Warden (GW), 30 Jan., 1935, TNA Acc. No. 16, 22/13, Vol. II; and 'Elephant Control in Tanganyika Territory,' Memorandum by Commander D.E. Blunt, 1933, TNA 21773.
19 Acting Game Warden to Chief Secretary, 20 Jan., 1934. TNA 21726.
20 Reporting in 1931, the GW observed that despite killing an average of 278 elephants per year in crop protection in Lindi Province, 'all elephant have increased considerably since 1925.' 'Elephant Census in Lindi Province 1931,' Acc. No. 16, 22/3 TNA.
21 District Officer (DO), Liwale to PC, Southern, 19 Mar. 1941. TNA Acc. No. 19, 22/13, Vol II.
22 DO, Liwale to PC, Southern, 19 Mar. 1941. TNA Acc. No. 19, 22/13, Vol II.
23 Minute by Administrative Secretary, Lamb, 22 Dec. 1943. TNA 31796.
24 'Note of discussion in Administrative Secretary's office on May 22nd, regarding proposed amendments of the boundaries of the Selous Game Reserve, consequent of the projected elimination of the Liwale Administrative District'. TNA Acc. No. 16, 22/13.
25 'Migrant Labour in Liwale' and 'Labour Migration among the Tanganyika Ngoni, Southern Province,' TNA Acc. No. 16 37/105.
26 The British government estimated that animals consistently consumed one-quarter to one-third of the territory's food crop production. Minute, illegible, 4 Aug., 1933. TNA 21726.

References

Agrawal, A. and Redford, K. 2009. Place, conservation, and displacement, *Conservation and Society*, vol. 7, no. 1, pp. 56–58.
Beachey, R. W. 1967. The East African ivory trade in the nineteenth century, *Journal of African History*, vol. 8, no. 2, pp. 269–290.
Benjaminsen, T. A. and Bryceson, I. 2012. Conservation, green/blue grabbing and accumulation by dispossession in Tanzania, *Journal of Peasant Studies*, vol. 39, no. 2, pp. 335–355.
Brockington, D. and Igoe, J. 2006. Eviction for conservation: a global overview, *Conservation & Society*, vol. 4, pp. 424–470.
Brockington, D. and Duffy, R. 2010. Capitalism and conservation: the production and reproduction of biodiversity conservation, *Antipode*, vol. 42, no. 3, pp. 469–484.
Brockington, D. and Scholfield, K. 2010. The conservationist mode of production and conservation NGOs in sub-Saharan Africa, *Antipode*, vol. 42, no. 3, pp. 551–575.

Büscher, B., Sullivan, S., Neves, K., Igoe, J. and Brockington, D. 2012. Towards a synthesized critique of neoliberal biodiversity conservation, *Capitalism Nature Socialism*, vol. 23, no. 2, pp. 4–30.

Corson, C. 2010. Shifting environmental governance in a neoliberal world: US aid for conservation, *Antipode*, vol. 42, no. 3, pp. 576–602.

Fairhead, J., Leach, M. and Scoones, I. 2012. Green grabbing: a new appropriation of nature?' *The Journal of Peasant Studies*, vol. 39, no. 2, pp. 237–261.

Garland, E. 2008. The elephant in the room: confronting the colonial character of wildlife conservation in Africa, *African Studies Review*, vol. 51, no. 3, pp. 51–74.

Giblin, J. and Monson, J. (eds). 2010. *Maji Maji: Lifting the fog of war*. Leiden and Boston, Brill.

Great Britain Colonial Office. 1923. *Report on Tanganyika Territory for the year 1922*. London, His Majesty's Stationery Office.

Great Britain Foreign Office. 1920a. *German Colonization*. London, His Majesty's Stationery Office.

Great Britain Foreign Office. 1920b. *Tanganyika (German East Africa)*. London, His Majesty's Stationery Office.

Hakansson, N. T. 2004. The human ecology of world systems in East Africa: the impact of the ivory trade, *Human Ecology*, vol. 32, no. 5, pp. 561–591.

Hakansson, N. T., Widgren, M. and Börjeson, L. 2008. Introduction: historical and regional perspectives on landscape transformations in northeastern Tanzania, 1850–2000, *International Journal of African Historical Studies*, vol. 41, no. 3, pp. 369–382.

Harvey, D. 2003. *The New Imperialism*. London, Oxford University Press.

Harvey, D. 2005. *A Brief History of Neoliberalism*. Oxford, UK, Oxford University Press.

Igoe, J., Neves, K. and Brockington, D. 2010. A spectacular eco-tour around the historic bloc: theorizing the convergence of biodiversity conservation and capitalist expansion, *Antipode*, vol. 42, no. 3, pp. 486–512.

Iliffe, J. 1967. *Tanganyika under German Rule, 1905–1912*. Cambridge, UK, Cambridge University Press.

Iliffe, J. 1979. *A Modern History of Tanganyika*. Cambridge, UK, Cambridge University Press.

Kelly, A. 2011. Conservation practice as primitive accumulation, *Journal of Peasant Studies*, vol. 38, no. 4, pp. 683–701.

Kjekshus, H. 1977. *Ecology Control and Economic Development in East African History: The case of Tanganyika, 1850–1950*. Berkeley, University of California Press.

Koponen, J. 1988. *People and Production in Late Precolonial Tanzania*. Monographs of the Finnish Society for Development Studies No 2, Helsinki, Finnish Society for Development Studies.

Larson, L. 2010. The Ngindo: exploring the center of the Maji Maji Rebellion, in J. Giblin and J. Monson (eds), *Maji Maji: Lifting the fog of war*. Leiden and Boston, Brill, pp. 71–116.

MacKenzie, J. 1988. *The Empire of Nature: Hunting, conservation, and British imperialism*. Manchester, UK, Manchester University Press.

Marx, K. 1976. *Capital: A critique of political economy, Volume One*. New York, Vintage Books.

Matzke, G. 1977. *Wildlife in Tanzanian Settlement Policy: The case of the Selous*, Syracuse, NY, Maxwell School of Citizenship and Public Affairs.

Neves, K. and Igoe, J. 2012. Uneven development and accumulation by dispossession in nature conservation: comparing recent trends in the Azores and Tanzania, *Tijdschrift voor Economische en Sociale Geografie*, vol. 103, no. 2, pp. 164–179.

Ojeda, D. 2012. Green pretexts: ecotourism, neoliberal conservation and land grabbing in Tayrona National Natural Park, Columbia, *Journal of Peasant Studies*, vol. 39, no. 2, pp. 357–375.

Redmond, P. 1975. Maji Maji in Ungoni: a reappraisal of existing historiography, *International Journal of African Historical Studies*, vol. 8, no. 3, pp. 407–424.

Rempel, R. 1998. Trade and transformation: participation in the ivory trade in late 19th-century East and Central Africa, *Canadian Journal of Development Studies*, vol. 19, no. 3, pp. 529–552.

Rodgers, W. A. 1976. Past Wangindo settlement in the eastern Selous Game Reserve, *Tanzania Notes and Records*, 77–78, June, pp. 21–25.

Sunseri, T. 2010. The war of the hunters: Maji Maji and the decline of the ivory trade, in J. Giblin and J. Monson (eds), *Maji Maji: Lifting the fog of war*. Leiden and Boston, Brill, pp. 117–148.

Tuck, M. 2009. Woodland commodities, global trade, and local struggles: the beeswax trade in British Tanzania, *Journal of Eastern African Studies*, vol. 3, no. 2, pp. 259–274.

Watts, M. 1983a. *Silent Violence: Food, Famine and Peasantry in Northern Nigeria*. Berkeley, University of California Press.

Watts, M. 1983b. On the poverty of theory: natural hazards in context, in K. Hewitt (ed.), *Interpretations of Calamity from the Viewpoint of Human Ecology*. Boston, MA, Allen & Unwin, pp. 231–262.

Watts, M. 1987. Drought, environment and food security: some reflections on peasants, pastoralists and commoditization in Dryland West Africa, in M. H. Glantz (ed.), *Drought and Hunger in Africa: Denying famine a future*. Cambridge, UK, Cambridge University Press, pp. 171–211.

8

Stopping the Serengeti Road

Social Media and the Discursive Politics of Conservation in Tanzania

Benjamin Gardner

Power, Meaning and Knowledge: Michael Watts and Political Ecology

During my second year of graduate school in UC Berkeley's Geography Department I took Michael Watt's political ecology seminar. On the first day of class he discussed relational power and how meaning making requires 'repertoires of knowledge'. To understand power relations, we needed to understand how those repertoires came into being, how power was 'built up, acquired and maintained through practice'. Understanding power was what we were all interested in and here was Professor Watts giving us the goods on day one. On that first day in class Watts was trying to show us that the material consequences and outcomes of struggles over resources were always simultaneously struggles over meaning. To 'do political ecology the Berkeley way' was to seek to understand the making of specific political economic relationships by understanding how struggles over meaning enabled certain repertories of knowledge to become sedimented as common sense. And how that knowledge influenced the value and meaning of material objects, places and people. The challenge for future scholars was precisely to try to understand and explain how embodied practice, worlds of meaning and power relations fused, merged and came into being at particular conjunctural moments.

Other Geographies: The Influences Of Michael Watts, First Edition. Edited by Sharad Chari, Susanne Freidberg, Vinay Gidwani, Jesse Ribot and Wendy Wolford.

That class helped give me the language and tools to embark on research into how power relations shaped the meanings and practices around conservation, land rights and tourism in Tanzania. This chapter is one attempt to grapple with the ways that conservation discourse is practised and embodied. I explore how sedimented repertoires of knowledge continue to structure political ecological relationships, sites of struggle and worlds of meaning.

Introduction

In June 2014 conservationists around the world celebrated a ruling by the East African Court of Justice [EACJ] that put a stop to plans by the government of Tanzania to build a paved highway across the Serengeti National Park. The ruling came almost four years after significant international opposition to the road project was initiated by several prominent NGOs, conservationists, governments, scientists and tourism companies. In 2010 the African Network for Animal Welfare (ANAW), a Kenyan-based animal rights NGO filed a lawsuit to halt the road construction. Serengeti Watch, an organization founded in 2010, provided funding for the case and helped mobilize the international community against the project. Serengeti Watch developed a social media strategy to 'rally world opinion, petition the government of Tanzania, bring attention to media within and without Tanzania, and enhance the work of other NGO's'.[1] One of their main strategies was to create a Facebook page 'Stop the Serengeti Highway', with the slogan, 'If we can't save the Serengeti what can we save?' Since that time Serengeti Watch and their social media campaign have played an important role in advocating against the road and educating the public on the issue.

There has been significant discussion and debate about the road largely framed in terms of conservation versus development. I am interested in how the debate over whether or not to build the road became an important event in itself, and a critical space to reproduce particular meanings of conservation in Tanzania.[2] In his study on safari tourism and representation, Andrew Norton (1996, 369) writes: 'safari tourism constructs East African nature through a process of nature *negotiation* between texts, tourists and places.' Norton's intervention was to go beyond textual analysis to incorporate the experiences of tourists as important sites of knowledge production about East African nature and landscapes. Cassie Hays (2012, 263) elaborates on this by illustrating how safari tourism, with mobility as a defining characteristic, is a performative practice that helps to construct the meaning of place. 'Nature on safari in Tanzania … is produced and consumed through a variety of touristic acts. ... Safari creates place through the process of categorization, in which participants see places as types rather than unique locations.' In this chapter, I argue that the opposition to the Serengeti road project provided an influential forum for a number of experts to reaffirm a narrative of the Serengeti as a singular place in need of protection from outside organizations and individuals.

Conservation is a relatively capacious idea, practice and set of places (that fit a certain type of place) in which different interests, ideas and values compete for a vision of how to manage nature and balance ecological, social and economic values (Adams and Hulme, 2001; Agrawal and Gibson, 1999; Anderson and Grove, 1987; Brockington 2004; Brockington and Duffy, 2010; Büscher 2010; Nelson, Gardner, Igoe and Williams, 2009; Ngoitiko, Sinandei, Meitaya and Nelson, 2010; Schroeder 2000; West 2006; Western 2001). The arguments and justifications used to oppose the building of the Serengeti highway help to present a relatively narrow vision of conservation that encourages a strict separation of people from nature. Such definitions of conservation, while still dominant, have been challenged over the past decade in Tanzania and other conservation hotspots around the world. National and local organizations and communities have either seized on ideas promoting community conservation, or drawn on discourses of sustainable land use practices to form coalitions promoting new understandings of conservation. Events such as the spectre of building a paved road through Serengeti National Park can serve to simplify ongoing social and political struggles over the meaning of conservation. In this case, one of the effects of the debate is to reproduce the idea of the Serengeti and its surrounding community lands as sites of authentic African nature (see Adams and McShane, 1992; Neumann 1997). This structurally undermines challenges by community groups who depend on a more open understanding of conservation that includes a strong role for local people. Contrasting a common-sense Western view of conservation with 'African' interests can help justify policies including the eviction of Maasai pastoralists from their lands in Loliondo in the name of this kind of conservation.

The international social media campaign against the road project helped to solidify the widespread belief among tourists and conservationists that the Serengeti is first and foremost a world heritage site whose value is measured at a global rather than a national or local scale. This discourse of the Serengeti as an authentic and universal site of nature that must be protected from the threats associated with development in general and specifically a paved road undermines competing discourses in which conservation and local land use interests can overlap. This conservation discourse hearkens back to more command and control or 'fortress conservation' ideas that help legitimate violence in the name of conservation.

This social media movement to defend the Serengeti from development reasserts forms of knowledge based on universal ideas of conservation and undermines emerging forms of knowledge by Tanzanians working in the trenches of community conservation. This social media campaign highlights the voices of scientists, conservationists, policy makers and tourists advocating for a vision of conservation that both wittingly and unwittingly justifies ongoing efforts to evict people living adjacent to the Serengeti National Park. The site is a curation of particular people, images and symbols that together tell a story of the Serengeti, what it is, what it means and who needs to protect it. When there are Tanzanian voices they are either represented as collective African wisdom, or as specific acts of protection by park rangers and managers. Local opposition to the road largely expressed concerns of the impact the road would have on immigration and land speculation. Local opposition was not anti-development in the way that the international movement framed the issue. Rather local leaders in Loliondo were concerned about a type of development brought by the road which would ultimately transform access to, and introduce alternative use values of, their land.

The first section of the chapter brings together textual and ethnographic analysis of the emergent discourse to stop the Serengeti road project and 'save the Serengeti'. The second part of the chapter draws on ethnographic research to demonstrate how this reinvigorated discourse underwrites violent evictions in Loliondo in the name of conservation in the Serengeti region. I focus my analysis on the ideas, discourses and examples used by Serengeti Watch and other influential conservation organizations to oppose the road project. I show how the threat of the impending road was important to consolidate a new commonsense position on conservation in the Serengeti region. The hegemonic position was contrasted with and then mobilized to help justify violent evictions and enclosures of Maasai in Loliondo.

The Serengeti Road

The idea of building a road through the Serengeti was first proposed over 20 years ago but failed to secure funding from the World Bank due both to environmental and monetary concerns. During the presidential campaign of 2005, then candidate J.M. Kikwete promised to build a highway through the Serengeti, connecting the northeastern part of the country with Lake Victoria and the more densely populated Western

Tanzania. There was little mention of the road project during Kikwete's first five-year term. On 17 May 2010 the President announced plans to build the road starting in 2012. The most controversial aspect of the proposed 385 km road was 'Natta-Mugumu-Tabora-Kleins Camp-Loliondo Road' (NMKL), a 54 km section through the northern part of Serengeti National Park (Gettleman 2010; Ihucha 2010; Ismail et al. 2010; Tanzania Natural Resource Forum Writers 2011).

Soon after the announcement two of the largest conservation organizations working in the Serengeti region, the Frankfurt Zoological Society and the African Wildlife Foundation, denounced the plan and proposed alternative road projects that would avoid traversing Serengeti National Park ('AWF Opposes Proposed Serengeti Highway' 2010; Frankfurt Zoological Society n.d.). On 6 June 2010 a group of United States-based conservationists started the 'Stop the Serengeti Highway' Facebook page and the 'savetheserengeti.org' website, followed by creating 'Serengeti Watch' a 'permanent non-profit organization' under the umbrella of the United States-based Earth Island Institute. The two directors of Serengeti Watch, Boyd Norton and David Blanton, are writers, photographers and conservationists who make a living by leading photographic safaris to East Africa as well as selling their photographs and books about African wildlife. I will return to this organization later in the chapter.

In July 2010 the Tanzanian President announced that the road would not be paved, which did little to quell the opposition. Donors such as the German government and the World Bank offered funding for one of the southern alternative routes. The President who remained committed to building the road despite the growing international opposition publicly rejected these offers. In December 2010 the Kenyan-based conservation organization, African Network for Animal Welfare (ANAW) filed an order in the East African Regional Court to stop the project based on ecological concerns. In April 2012, the Chinese company which was to build the Musoma to Tanga railroad conducted an environmental impact assessment. In August 2013 the ANAW trial was heard by the EACJ with no immediate ruling. In September 2013, local contractors began working on roads on the western side of Serengeti. On 20 June 2014 the EACJ ruled for ANAW claiming that building a 'Bituman' or paved road would be 'illegal'. Conservation groups were relieved but still concerned that Tanzania would build a gravel road in its place. In June 2014 the Tanzanian government approached the German government to proceed with a feasibility study for the alternative southern route. During the four years of proceedings from June 2010 to June 2014 there was a robust international response to the possibility of the road.

'Worldwide Opposition' and Scientific Expertise

An article from the African Wildlife Foundation [AWF] on 15 March 2011 claimed: 'Worldwide opposition to road has yet to sway Tanzanian government.' The article stated: 'What began with science-based opposition by AWF and other conservation groups has grown into a grassroots movement playing out across social media outlets and echoed by global influencers such as the Work Bank, UNESCO, and … the German Government. The diverse coalition of scientists and conservationists, economists and development experts has coalesced around a singular message: The construction of a road through Serengeti National Park would gravely threaten the last great migration of hundreds of thousands of wildebeest and erode one of the most pristine landscapes on earth' ('Worldwide Opposition to Serengeti Road Has Yet to Sway Tanzanian Government' 2011, 1). AWF has a large presence in northern Tanzania and is deeply invested in the national park model of conservation in the country. There is nothing surprising about their stance or their effort to marshal scientific and grassroots support to protect African wildlife habitat. Yet, the controversy provided a platform for AWF to represent an international point of view of conservation that they claimed represented 'worldwide opposition'.

One of the most compelling arguments used by AWF and other conservation organizations was to cite the opinion piece, 'Road will ruin Serengeti' published in September 2010 by 27 scientists in *Nature* magazine (Dobson et al. 2010). The group included prominent scientists and conservationists, among them Andrew Dobson an ecologist from Princeton University, Markus Borner the director of the Africa Programme at Frankfurt Zoological Society and Anthony Sinclair from the Center for Biodiversity Research at the University of British Columbia. The two-page piece claimed: 'the road will cause an environmental disaster' and that based on evidence from other parts of the world was likely to cause 'the Serengeti ecosystem to collapse, and even flip from being a carbon sink into a major source of atmospheric carbon dioxide'. The authors highlighted the Serengeti's rare and iconic ecosystem and that 'classic, long-term studies there have made fundamental contributions to knowledge of how natural ecosystems function'. They suggested that fences would be needed to protect wildlife from vehicles 'as happened in Banff National Park in Canada' and would fragment the ecosystem. They claimed that this decision would cause a disastrous domino effect. The authors stated: 'simulations suggest that if wildebeest access to the Mara River in Kenya is blocked, the population will fall to less than 300,000 [from 1.3 million]. This would lead to more grass fires, which would further diminish the quality of grazing by volatizing minerals, and the ecosystem could flip into being a source of atmospheric CO2. ... There would be far fewer game, fewer predators and more than 80% of the park would burn every year.'

The authors concluded by calling on the Tanzanian government to choose a better way by exploring a southern road alternative, which would go around the southern end of the park. A short rebuttal by social scientists Katherine Homewood, Daniel Brockington and Slan Sullivan, claimed the authors had overstated their case. They also pointed out what few people were talking about, that although 'wildlife tourism is a major contributor to Tanzania's economy ... local people whose land-use practices have created and now maintain the ecosystems valued by Western conservationists, are being dispossessed by state and international elites who capture much of the tourism revenue and reinvest it in conservation-incompatible land use' (Homewood, Brockington and Sullivan, 2010). Despite this pointed critique there was little opposition to the growing consensus among scientists.

Three years after the *Nature* article was published biologists from the Norwegian University of Science and Technology, led by E. Røskaft came to a different conclusion on the ecological impacts of the road. The scientists concluded that the road was not as great a threat as the authors of the *Nature* essay had indicated, and that there were greater threats to the ecosystem. They claimed that 'road building ranks far lower as a threat than issues such as climate change, poverty, high population densities and deforestation, particularly north of the park in Kenya' (Fyumagwa et al. 2013). Despite this significant scientific challenge, these studies were not cited or circulated widely in the popular conservation media and social media. By the time these studies were published the idea that there was a worldwide consensus had already gained significant momentum and new scientific evidence did little to slow its influence.

'If we can't save the Serengeti, what can we save?'

Not only is conservation a way of thinking about economic and political values, it is also big business. Organizations like AWF and FZS (Africa Programme) exist because of African wildlife and national parks. The scientists who published the *Nature* opinion piece also make their living by doing research about African wildlife and conservation. As a researcher and professor I too am part of this international political economy of African wildlife and conservation. Social media is frequently framed as existing outside of these cultural and economic relationships. The authority and authenticity of conservation groups is derived in part from the perception that they do not have specific monetary interest in the issue.

In 2010 two American-born conservationists Boyd Norton and David Blanton founded Serengeti Watch, a non-profit organization under the umbrella of the 501(c)3 conservation organization Earth Island Institute. The organization was founded in response to the road proposal. Its mission was 'to build a coalition of support, advocacy, and funding for the Serengeti ecosystem, the people living around it, and adjacent reserves and protected areas'. Its slogan, 'if we can't save the Serengeti, what can we save' serves as a call to action by the global conservation community. Saving the Serengeti is posed as a litmus test for the transnational environmental movement.

As with many environmental causes the leaders are portrayed as answering an ethical call to action. Not only do Boyd Norton and David Blanton have significant credibility as experts in the African conservation community (which exists largely outside of Africa and is dominated by non-Africans), they both make a considerable amount of money from their photography, writing and speaking about African conservation, as well as guiding safaris. The advisory board for Serengeti Watch consists of one person, Jim Fowler, who served as Marlin Perkins' co-host and eventually the sole host of the well-known Mutual of Omaha's *Wild Kingdom*. Together these three men represent a recognizable public face of African conservation in the United States and Europe.

While there were many groups and individuals involved in the debate over building a paved road across Serengeti National Park, Serengeti Watch and its Facebook page played a particularly important role among tourists who had either visited Tanzania or taken an interest in African wildlife and conservation.[3] The Facebook page largely focused on the perils of building the road and advocated for alternatives. But over the past four years of debate it also became a more all-purpose space to advocate for wildlife protection and conservation, with a particular slant toward animal welfare and animal rights in Africa.[4]

The Facebook page is quite robust with an average of three to four posts per week between 2010 and 2014. Many of the posts are links to articles about conservation, statements from prominent biologists or ecologists, and quotes from famous conservationists such as Jane Goodall. These more evidence-based scientific articles are mixed with more philosophical and moral statements about the meaning and value of African wildlife. For example, one of the early posts on 13 October 2010 was by Duke University Professor Anne Pusey who, according to the post, studied lions in the Serengeti for 10 years. She wrote: 'The Serengeti is a unique and precious ecosystem – one of the very few large scale migratory systems of large animals remaining on the planet ... A road across the migratory routes will devastate the system for all the reasons listed in this [petition] letter and survey.'[5] This statement was rather typical of accounts by scientists who are experts on African wildlife that appeared regularly on the site. The following day, 14 October 2010, the organizers of the site posted this quote from American environmentalist Michael Frome: 'Wilderness, above all its definitions and uses, is sacred space, with sacred powers, the heart of a moral world.' The Facebook page mixed news stories, scientific papers and inspirational quotes to, as they say, 'connect world supporters, travelers, tour companies, local communities to raise awareness on important issues facing the Serengeti'.[6] Key themes on the site included conservation as a reflection of global social values, wildlife protection as a proxy for humanitarianism, and a reverence for scientists with fieldwork experience in East Africa.

I first learned about the Serengeti Watch while researching the proposed eviction of Maasai people from their lands in Loliondo, an area of Maasai villages bordering Serengeti National Park. Maasai communities in Loliondo had been fighting with international conservation organizations and Tanzanian government agencies over their land rights since a 2009 eviction. Enforcing the hunting rights of the Ortello Business Corporation (OBC), which held the trophy-hunting lease to the area since 1992 when police forcibly removed Maasai from their village land that overlapped with the hunting area. Although there were conflicts between the company and the community from the beginning, July 2009 was the most forceful and violent confrontation between state officials and the Maasai. Maasai leaders mobilized their own press and social media response and were able to stop the evictions after 48 hours. Maasai leaders and organizations did significant organizing at the local and regional level, while also reaching out to national and international allies to spread the word and put pressure on the Tanzanian government. Since that time Maasai leaders and NGOs have been embroiled in a conflict with state agencies about a proposed plan to create a 1500 km2 hunting reserve that would dispossess the Maasai of roughly one-third of their land. This is where the movement to stop the Serengeti road and the effort to create a new game reserve and evict the Maasai converged.

The Knowledge/Practice of Conservation

In November 2014, the Minister of Natural Resources and Tourism announced plans to demarcate Maasai village land as a new game reserve for hunting. This was the third time the government had proposed the idea. At each turn Maasai leaders and advocates met the announcements with immediate resistance. Central to the Minister's argument was the well-worn narrative that to protect nature effectively an external agency was needed to separate people and animals to prevent conflict. The Minister claimed that the local population was growing too large and presented a direct threat to wildlife. Population pressure had figured largely in the Tanzanian government's explanation of why they evicted Maasai from their land in 2009. Although population can play a role in sustainable resource management, a general belief that population growth poses an existential threat to nature and natural resources is often mobilized with little attention to local political economic and even ecological conditions. The belief that population growth is the largest threat to wildlife conservation continues to hold significant sway in the conservation communities

In an official response to the Loliondo conflict, then Minister of Natural Resources and Tourism Khamis Kagasheki explained why population pressure led him to the decision to create a new game reserve and separate people and wildlife (Kagasheki 2013). Maasai leaders challenged each of these claims saying that they managed the land adjacent to the Serengeti Park for both livestock and wildlife for centuries and there was scant evidence to support the belief that their eviction would protect wildlife. They largely saw this as an opportunity to appease an important foreign investor and fulfil a longstanding desire to extend the boundaries of Serengeti National Park. Maasai leaders attending a protest in 2013 flipped the government's overpopulation script.

The news of this attempt to create a new game reserve area and relocate the Maasai residents was reported and discussed on the Stop the Serengeti Highway Facebook page. Despite more specialized online forums like *Safari Talk* and *The Hunting Report*, the Facebook page had become a popular space for people interested in all things relating to the Serengeti. News of the new game reserve and the ensuing debate about the evictions fit neatly into the ongoing conversation about saving the Serengeti.

One of the followers of the site linked the effort to stop the road project with the effort to create the new game reserve, including the impending eviction. He spoke up on behalf of conservationists, or 'those of us who fight like hell to prevent the extinction of wildlife', and supported the government's efforts to expand conservation areas:

> Yes, we disagree with Tanzania's plan for the highway. However, an example of Tanzania attempting to protect wildlife habitat from over development is the controversial relocation of thousands of Maasai from Loliondo. When Maasai were initially permitted to remain in Loliondo, the population was less than 5,000 to now it is 40,000 to 60,000. The population explosion and permanent structures (not legally allowed) has destroyed the natural habitat, restricted movement of the migration and effectively eradicated big cats from major portions of the ecosystem. The relocation plan offers incentives (financial, free land, permanent water, education for

their children, and livestock). ... When Tanzania attempts to do something positive for the wildlife habitat, those of us who fight like hell to prevent the extinction of wildlife should support it.

It would be easy to dismiss one comment on a discussion board. But after reading thousands of posts, the more significant this comment and the linking of the two events appear. The Stop the Serengeti Highway Facebook page has over 60 000 followers. It is through posts and discussions like this that people often form their opinions and take a stand on issues. The author of the post, William Cowger, is an American-born conservationist and photographer in the mould of Boyd Norton and David Blanton. His website and Facebook page promote his Safari company, William Cowger's Safaris, where you can purchase his wildlife photographs or join him on an upcoming safari. Similar to Norton and Blanton, Cowger drew on his personal experience to represent the Serengeti. The social media site amplified his status as an expert and positions him as someone willing to tell hard truths to others who perhaps had never visited the Serengeti or known it in the same ways as he did. Personal experience, especially by white Western conservationists, is revered in the conservation community. I am neither singling out nor targeting Cowger as a unique example of hubris and exaggeration. I think it important to take him seriously. Debates about the meaning of African wildlife, their value and the policy options available to governments and NGOs influence many people who are involved in international conservation.

It is fair to say that Cowger benefits from African conservation. His photographs, safari tours and identity as a lay expert on African wildlife bring different benefits and value to him. The Facebook page with an eager and committed audience only heightens his status to represent what is in the best interest of the Serengeti (and create the impression that this place has a discrete set of self-interests). It is fair to say that there are many problems in Cowger's posts. He cites little to no research, historical precedent or even facts in his analysis that relocating the Maasai is both necessary and fair. Cowger says that 'when the Maasai were allowed to remain in Loliondo' they were far fewer. He seems to be citing the Minister's press statement to determine the land rights status of the Maasai. For anyone following this issue closely, the press release can hardly be seen as an unbiased assessment.

The Serengeti Shall not Die

The original movement to 'save the Serengeti' was itself born from a Western fascination with and interest in African nature and wildlife. Rod Neumann (1998) documents how the process to advocate for the protection of the Serengeti began with conservation societies in Europe in the 1930s and 1940s. With foreign interest in wildlife protection in Tanzania intensifying, the most prominent figure promoting the creation of Serengeti National Park was the German veterinary professor and Frankfurt Zoo director, Bernhard Grzmek. Historian Thomas Lekan explores Grzmek's tactics and motivations promoting the protection of Serengeti National Park. Like other Western conservationists at the time Grzmek sought to find and save an authentic piece of nature. To promote his work and create a global constituency he produced two films, the 1956 documentary *No Room for Wild Animals* and the 1961 documentary *Serengeti Shall Not Die*. Similar to other iconic wildlife television programmes like Mutual of Omaha's *Wild Kingdom* in the United States, Grzmek's films and German television show *A Place for Animals* helped introduce faraway audiences to the beauty and value of African wildlife. 'Grzmek is arguably [the] most influential publicist for African wildlife in postwar Europe, though his legacy is more contested.' As one observer has noted, in his thirty-year career Grzmek 'probably raised more money for conservation, educated more people about nature, and twisted more arms of more African bureaucrats than any man in history' (Lekan 2011, 227).

Like the Serengeti Watch Facebook page Grzmek's writings and films mixed ecology with moralistic proclamations. 'By presenting scientific research as heroic adventure, *Serengeti Shall Not Die* drew upon and reworked a host of tropes and images of African wilderness as both a "lost paradise" awaiting cultivation and a "green hell" of violent natives and tropical disease-which had long fired the European imaginations' (Lekan 2011, 232).

Grzmek's films and writings told stories where protecting African nature was seen as the truest sign of humanity in the face of modernizing forces that would soon transform the entire earth into fields, farms and cities. Grzmek implored his audiences in Germany and across Europe to help save African wildlife from humans. The human intrusion he spoke of was both general, a type of Western greed and capitalist expansion, and local, peasants and pastoralists set on taming the land according to their own backward desires. Grzmek believed that Maasai pastoralists would eventually destroy the Serengeti grasslands, arguing that population growth would spell the end of African nature as we knew it. 'You cannot keep men, even black and brown ones, from multiplying and cannot force them to remain "primitive"' (Lekan 2011, 238).

In this tradition, Cowger too claimed that population has destroyed habitat and restricted the migration. Despite limited evidence to support such claims he repeats the well-worn argument also mobilized by the Minister. Dominant discourses such as the Malthusian narrative of population 'explosions' travel and gain traction by being used by different people in different spaces (Ross 1998). Conservation organizations have increasingly sought out pastoralist areas, recognizing that overall land use is often more important than population. Despite changing understanding of pastoral land use, Cowger simply repeats the highly legible phrases linking population and destruction.

Cowger ends his post claiming that evicting the Maasai is a necessary action and conservationists should support it. In a later post he says: 'I am not choosing sides, just portraying both sides.' In one sentence, he absolves himself of any accountability, while affirming his expertise on the topic. In many ways Cowger is an example of a member of the public who is able to become an expert on African conservation through social media. His credentials as a wildlife photographer and tour guide give status and legitimacy, especially among members of the public that are relatively new to the specifics of this issue.

Conclusion

The circulation of information, the production of knowledge and the creation of meaning are essential elements of conservation practice in Tanzania. As Peluso and Watts (2001, 6–7) contend, 'violence stands awkwardly in respect to environmental concerns. The environment is increasingly present and yet frequently hidden by both the perpetrators and observers of violence alike.' The seemingly benign conversation in support of the Serengeti has wide ranging effects, including providing the discursive support for violent evictions and enclosures. The meaning of conservation is not fixed, but it conveys collectively agreed upon values and interests. What conservation is, what it can be, is a contested space, especially in a place like Tanzania where the different meanings of conservation can lead to collaboration or conflict. In this chapter I have shown how a social media campaign to stop the construction of a road through Serengeti National Park is a site of knowledge production and meaning making. Social media campaigns like Serengeti Watch are important means for organizing an understanding of the African landscape, why it matters, what saving it means and who is needed to protect it.

Notes

1 http://www.savetheserengeti.org/about-us/ (accessed 1 September 2015).
2 The discussion and debate over the road is itself an intervention in the understandings of African wildlife and landscapes.
3 I focus my analysis on this particular social media campaign as it reached a broad audience of people interested in African wildlife and conservation.

4   The different perspectives over the role of animal rights in African conserva-
    tion were highlighted in the Western media when an American dentist killed
    a beloved Zimbabwean Lion, named Cecil. The ensuing coverage over the
    killing and hunting of African wildlife in general revealed significant divides
    in the conservation community over the role of hunting for conservation. It
    is important to remember that ANAW, a staunch anti-hunting organization,
    was highly influential in this particular campaign against the building of the
    road.
5   She is referring to the petition started by Serengeti Watch, 'Highway
    Development Threatens Serengeti'. http://www.savetheserengeti.org/issues/
    highway/stop-the-serengeti-highway/ (accessed 1 September 2015).
6   About Serengeti Watch. http://www.savetheserengeti.org/about-us/ (accessed
    1 September 2015).

References

Adams, J. S. and McShane, T. O. 1992. *The Myth of Wild Africa: Conservation
without illusion*. New York, Norton.

Adams, W. M. and Hulme, D. 2001. If community conservation is the answer in
africa, what is the question? *Oryx*, vol. 35, no. 3, pp. 193–200.

African Wildlife Foundation. n.d. 'Proposed Serengeti Highway' (Position
Statement), African Wildlife Foundation.

African Wildlife Foundation. 2010. 'AWF opposes proposed Serengeti Highway'
(African Wildlife Foundation Newsletter). https://www.awf.org/news/awf-
opposes-proposed-serengeti-highway (accessed 15 May 2017).

Agrawal, A. and Gibson, C. 1999. Community in conservation: beyond enchant-
ment and disenchantment, *World Development*, vol. 27, no. 4, pp. 629–649.

Anderson, D. and Grove, R. H. 1987. *Conservation in Africa: Peoples, policies
and practice*. Cambridge, UK, Cambridge University Press.

Brockington, D. 2004. Community conservation, inequality and injustice: myths
of power in protected area management, *Conservation and Society*, vol. 2, no.
2, pp. 411–432.

Brockington, D. and Duffy, R. 2010. Capitalism and conservation: the production and reproduction of biodiversity conservation, *Antipode*, vol. 42, no. 3,
pp. 469–484.

Büscher, B. 2010. Derivative nature: interrogating the value of conservation
in boundless southern Africa, *Third World Quarterly*, vol. 31, no. 2, pp.
259–276.

Dobson, A. P., Borner, M., Sinclair, A. R., Hudson, P. J., Anderson, T. M.,
Bigurube, G., and Wolanski, E. 2010. Road will ruin Serengeti, *Nature*, vol.
467, no. 7313, pp. 272–273.

Frankfurt Zoological Society. n.d. The Serengeti North Road project: a commercial road through Serengeti National Park jeopardizes the integrity of a
World Heritage Site. Alternative routes could easily meet the economic needs
and even improve the conservation status of Serengeti National Park.
(Presentation).

Fyumagwa, R., Gereta, E., Hassan, S., Kideghesho, J. R., Kohi, E. M., Keyyu, J. and Røskaft, E. 2013. Roads as a threat to the Serengeti ecosystem, *Conservation Biology: The Journal of the Society for Conservation Biology*, vol. 27, no. 5, pp. 1122–1125.

Gettleman, J. 2010. Serengeti road plan offers prospects and fears, *The New York Times*, 30 October.

Hays, C. M. 2012. Placing nature(s) on safari, *Tourist Studies*, vol. 12, no. 3, pp. 250–267.

Homewood, K., Brockington, D. and Sullivan, S. 2010. Alternative view of Serengeti road, *Nature*, vol. 467, no. 7317, pp. 788–789.

Ihucha. 2010. Serengeti highway to go ahead—Kikwete, *The East African*, 16 August.

Ismail, F. A., Kweka, Issack, E., Madayi, Z., Kuhanwa, Z. and Msuha, M. 2010. *Consultancy services for detailed engineering design, environmental and social impact assessment, and preparation of tender documents for upgrading of natta-mugumu-loliondo (171.9KM) road to bitumen standard, Tanzania* (Environmental and Social Impact Assessment (ESIA) Draft Report).

Kagasheki, K. 2013. Tanzania Government Proclaim About Loliondo Epic (Press Release). http://community.co.tz/2013/04/08/tanzania-government-proclaim-about-loliondo-epic/ (accessed 15 May 2017).

Lekan, T. 2011. Serengeti shall not die: Bernhard Grzimek, wildlife film, and the making of a tourist landscape in East Africa, *German History*, vol. 29, no. 2, pp. 224–264.

Nelson, F., Gardner, B., Igoe, J. and Williams, A. 2009. Community-based conservation and Maasai livelihoods in Tanzania, in K. Homewood, P. Kristjanson and P. C. Trench (eds), *Staying Maasai? Livelihoods, conservation and development in East African rangelands*. New York, Springer Science + Business Media, pp. 299–333.

Neumann, R. 1997. Primitive ideas: protected area buffer zones and the politics of land in Africa, *Development and Change*, vol. 28, no. 3, pp. 559–582.

Neumann, R. P. 1998. *Imposing Wilderness: Struggles over livelihood and nature preservation in Africa*. Berkeley, University of California Press.

Ngoitiko, M., Sinandei, M., Meitaya, P. and Nelson, F. 2010. Pastoral activists: negotiating power imbalances in the Tanzanian Serengeti, in F. Nelson (ed.), *Community Rights, Conservation and Contested Land: The politics of natural resource governance in Africa*. London, Earthscan/James and James.

Norton, A. 1996. Experiencing nature: the reproduction of environmental discourse through safari tourism in East Africa, *Geoforum*, vol. 27, no. 3, pp. 355–373.

Peluso, N. and Watts, M. 2001. *Violent Environments*. Ithaca, NY, Cornell University Press.

Ross, E. B. 1998. *The Malthus Factor: Poverty, politics and population in capitalist development*. London: Zed Books.

Schroeder, R. A. 2000. Beyond distributive justice: resource extraction and environmental justice in the tropics, in C. Zerner (ed.), *People, Plants and Justice: The politics of nature conservation*. New York, Columbia University Press, pp. 52–66.

Tanzania Natural Resource Forum Writers. 2011. The Serengeti highway controversy, *Swara*, April–June 2011.

West, P. 2006. *Conservation is our Government Now: The politics of ecology in papua new guinea*. Durham, NC: Duke University Press.

Western, D. 2001. Taking the broad view of conservation – a response to Adams and Hulme, *Oryx*, vol. 35, no. 3, pp. 201–203.

Worldwide Opposition to Serengeti Road has yet to sway Tanzanian Government. 2011. (African Wildlife Foundation Newsletter). https://www.awf.org/news/worldwide-opposition-serengeti-road-has-yet-sway-tanzanian-government (accessed 15 May 2017).

9

Privatize Everything, Certify Everywhere

Academic Assessment and Value Transfers

Tad Mutersbaugh

Certify Everywhere

> 'Certified everywhere ain't got to print my resume.'
> Migos, *Fight Night*

Certification provides an *everywhere* to privatization's *everything*, stabilizing the transfer of value via practices such as inspections and quality seals. Inspired by Michael Watts' (1994) essay 'The Privatization of Everything?', this chapter locates academic assessment in the context of efforts to standardize aspects of post-secondary education in a manner that facilitates a privatization of student finances and university services. While there may be a certain lure to being 'certified everywhere', captured niftily by Migo's 'resume' lyric with its promise of being everywhere recognized and never subjected to verification – an idealism likewise on display in the ISO World Standards Day 2015 poster (Figure 9.1), this chapter will rather argue that this 'common language' forms part of the 'semiotic machine' that 'naturalizes' social exploitation (Watts 2001).

This chapter examines *assessment*, an extension of standards-based accreditation and certification into academic life.[1] Academic assessment forms a practice linked to the accreditation of US universities by regional agencies, one that qualifies professors within them to confer degrees. Recently, as regional bodies have pressed universities to standardize the practices through which degrees are produced – such as specifying the

Other Geographies: The Influences Of Michael Watts, First Edition. Edited by Sharad Chari, Susanne Freidberg, Vinay Gidwani, Jesse Ribot and Wendy Wolford.

Figure 9.1 ISO-sponsored World Standards Day 2015 Poster: 'Standards are the World's Common Language'

hours of class time required for a credit unit and the use of student learning outcomes (SLOs) – assessment has increasingly incorporated certification activities such as syllabi inspection and the creation of equivalencies for inter-institutional credit transfers. This introduction of certification practices is linked to neoliberal privatization within academe insofar as a standardization of 'properties' (e.g. credit units) serves to regularize connections between the private and public spheres and to facilitate the introduction of privatized management services such as Blackboard®, TES® (Transfer Evaluation System) and private student lending. In sum, as in other forms of public asset privatization, certification and accreditation play two roles: they codify, equate and monitor value so that it may be shifted to new sites, and introduce a 'transparency' built on inspectibility that 'sanitizes' 'structural power' by incorporating it into the standards themselves (Dolan 2008, 310: see also Hudson and Hudson 2003; Mutersbaugh 2005).

In this chapter I apply an analysis inspired by Watts' triangulation of political economies of economic accumulation, discursive practices and everyday cultural practice. However, in place of his focus on resource extraction, for example of oil's 'pipelines and pumping houses', this chapter treats educational assessment practices such as credit transfers and student learning outcomes. I argue that these educational practices work in tandem, creating equivalencies that in turn allow value to be 'piped' (to extend the oil metaphor) from 'non-branded' state schools and for-profit institutions to 'branded', 'destination education' universities – whether elite research institutions like Berkeley or sports powerhouses like Kentucky's 'Big Blue Nation'. I further argue that these value transfers are linked to an increased use of casualized labour that is monitored through assessment.

Although this chapter's principal aim is analytical, I hope that it may also contribute to 'slow scholarship' strategies by highlighting the role that assessment-promoted standards play in shaping the labour landscape of neoliberal universities – which now includes a great many assessment-related staff and extra-institutional liaisons in addition to the professoriate and administrators (Mitchell 2003; Davies and Bansel 2010; Ball 2012; Slaughter and Rhoades 2000). Calls for a grassroots 'slow scholarship' movement to counter the demoralizing and counterproductive effects of academic speed-up properly targets the introduction of an 'audit culture' into academe and calls for labour-organizing efforts (Mountz et al. 2015; see also Mott, Zupan, Debbane and R 2015). In examining assessment, we may develop a better understanding of those dedicated to managing the academic assembly line (Brown 2016).

Pipes and Tubes

To begin this analytical triangulation, a focus on the role of 'pipelines and pumping stations' – which is to say on the political economic mechanisms that mediate value transfer and the discourses supporting this transfer – may help to conceptualize academic commodification. Replacing 'oil' with 'academia', we might agree that '[m]any of those who write about oil typically, and rather curiously, have little to say about the materiality ... and the political economy of what falls within the circumference of a vast, complex industry' (Watts 2012, 439). In applying Watts' resource-economic concepts of value transfer, territories, frontiers and violence to educational assessment, clear differences emerge: student credit hours attach to and 'qualify' individuals in a manner that oil cannot, and unlike the transnational ambit of petroleum products, 'frontiers' and 'territories' mark spatial and structural divisions between and within a university system often governed under a single authority. Practices of commodification may differ, yet as public education has become a target of initiatives designed to extract public resources, Watts' analysis of rentier political economy provides a useful guide for parallel critique.

Transferring Value Across Frontiers

The (neoliberal) vision of 'a common language' signals the role of standards as a mechanism to 'pipe' value across '... frontiers, understood not simply as a territory peripheral to, or at the margins of the state ... but as a particular space – at once political, economic, cultural, and social – in which the conditions for a new phase of ... accumulation are being put in place ...' (Watts 2012, 445). As with oil, academia's 'new phase' of accumulation is built in part through a transfer of surplus value across space – and, as with oil, the value of the transferred commodity must be assured. For universities, 'credit units' are one such commodity: their transfer requires agreements regarding the constitution of a 'unit' of value and also the use of varied semiotic and labour processual (performative) practices to stabilize these units as they travel within and between universities. Practices used to stabilize inter-mural transfer include the elaboration of 'common core' curricula and SLOs (Student Learning Outcomes). Common core curricula specify that undergraduate associate degrees earned in one school may be transferred to another, and that each of the specific courses within those degrees will satisfy their correlate course requirement in a university to which a student may transfer. Starting with the first two legs of Watts' analytical triangulation noted above, this section and the next one examine first the rentier relations that inhere in credit units and SLOs and then the discursive practices used to justify and stabilize these political economies. A subsequent section takes up the last leg, showing how assessors (assessment staff) perform 'emotional labour' as they work to integrate professors into assessment activities.

Credit unit transfers and SLOs each create and assign a sort of market 'equivalence', and each is promoted by conservative educational policy foundations such as the Educational Commission of the States. These organizations have successfully pushed for legislation in support of credit unit equivalence, a 'top-down' strategy which helps ensure that any course taken in one university will count for a similarly named course in any other university.[2] SLOs, also supported by financial services-linked educational policy foundations such as Lumina,[3] represent the 'bottom-up' efforts of university assessment officers to create national rubrics for the learning outcomes of these similar courses, Though distinct initiatives, credit unit transfers and SLOs work together to facilitate value flows: legally binding 'transfers' policies limit a university's ability to scrutinize course credit transfers, while SLOs seek to specify – and at a national level standardize – course content and presentation.

By permitting value transfer, these standardization practices also extend the 'where' of academia, creating frontiers where value may be appropriated, and that, to follow Watts' characterization of resource extraction, may be 'seen as spaces ... in which violence and political negotiation [are]...at the center of social and economic life' (2012, 446). The academic spaces that have been reworked to serve as frontiers for value extraction include underfunded community and state colleges, many of which subsidize the cost of producing a credit unit through reliance on an increasingly casualized, adjunct faculty labour force. This 'harmonization' of credit hours – to use the certification lexicon – allows credits earned at these poorer colleges to be marked as 'equivalencies', and therefore to count at branded 'flagship' institutions where credit hours may cost anywhere from twice to five times as much (College Board 2015).[4]

This raises the question of who benefits. Clearly the professoriate as a whole loses access to better-paid, tenured job opportunities that are replaced by temporary adjunct positions. At the same time, the management of equivalencies creates jobs for assessment personnel. Do students benefit? The case may be made that cheaper credit hours benefit students channelled into two-year and non-branded institutions – particularly students of colour who are disproportionately represented in two-year institutions (Aud, Fox and KewalRamani 2010). If education is constituted as specific skills and knowledge dictated in SLOs, then certainly a lower dollar cost paid per skill acquired would benefit students. However, if education is understood to be a relational activity in which students learn to collaborate in knowledge production, then their interactions with a casualized, adjunct faculty typified by high turnover rates – currently over 70% of US faculty are transient, with 51% part-time and 19% non-tenure-track[5] – might be less satisfactory. Again, this particularly affects students in two-year institutions where the percentage of contingent faculty is relatively higher (AFT 2010). In any case, the national effort underway to enshrine transfers in state law leaves branded universities and their faculty with little say regarding the teaching conditions under which credit units are produced. Meanwhile, the creation of national SLOs serves to provide the appearance that the content is equivalent.

Semiotics of Rent Extraction

> *'[I]t is surprising how little work has focused on the invention of institutions which produce, transmit and stabilize development "truths".'*
>
> Watts (1993, 263)

Alongside pipes, tubes and credit hours, an assessment 'economy of semiology' serves to naturalize educational practice as '... routinized exchanges of codified information' (Dolan 2008, 309). A second vertex of Watts' triangulation explores the role played by familiar neoliberal tropes in stabilizing the practices of resource extraction through the creation of 'truths' that justify, even necessitate, the use of coercion and tacit acceptance of surveillance. One discourse common to resource extraction economies that may (if perhaps incongruously) be applied to educational practice, relies on a 'vocabulary of risk, [that feeds] the great semiotic machine that naturalizes the consequences of social exploitation' (Watts 2001, 139).

Risk in resource extraction includes not only theft and contemporary market fluctuations, but also longer-term constraints on value growth that would be occasioned by global climate change initiatives. Educational assessment also contains a risk vocabulary, such as in the instance of 'curricular drift'. During recent efforts by University of Kentucky administrators to introduce a new 'common core' curriculum in line with national private foundation initiatives (e.g. Lumina), supporters argued that professors who taught courses under the previous 'general education' curriculum had allowed syllabi to 'drift' from the original curricular intent. This wayward curriculum thus needed replacing by the new 'common core' curriculum, which would itself require 'continuous' assessment to mitigate risk. According to this narrative, ongoing course monitoring activities would hold professors accountable and prevent a future drift. Proponents of assessment provided no data or studies to support their claims, yet nonetheless received the (previous) Arts and Sciences Dean's support. Assessment since implemented is linked both to state legislated routinized transfer agreements and SACSCOC (Southern Association of Colleges and Schools Commission on Colleges) accreditation procedures.

As with other instances of standards-based certification, assessment mechanisms serve to govern and shift risk to less powerful 'value chain' members (Otto and Mutersbaugh 2015). Standards-based assessment works with potential sources of 'risk' (e.g. student retention) to establish protocols to control these risks and shift both the labour of risk management *and* the consequences of failure to professors. As standardized 'core' curriculum and the associated credit transfers become routinized, rent extraction (in cash, labour and students' personal information)

Figure 9.2 ISO video – Standards open up world markets. http://www. youtube.com/user/worldstandardscoop[6]

becomes less visible and 'stabilized' – as suggested in Watts' quote (and see Figure 9.2). The question inspired by this analysis of educational discourses is that of precisely how the work of the professoriate contributes to a 'naturalizing' of exploitation, and why the narrative of curricular drift seemed to resonate with professors. One confusion may attach to the discursive resonances that link neoliberal rentier initiatives such as 'common core' curricula with the pre-existing and remarkably persistent liberal university's 'imperial' knowledge model, which supports the perception among branded university professoriates that they should play a role in setting the character of 'common' curricula (Spivak 1993; Andreotti 2006; Harney and Moten 2013; Mott et al. 2015). When a flagship institution is provided with the opportunity to design the common core curriculum that will be subsequently taken up by two-year and other state institutions, it seems to conform to the liberal model of knowledge production, even though the neoliberal political economy of the common core involves an intensified exploitation of adjunct faculty in other state institutions. A second confusion, which I take up in the final section, arises in our experience of the university labour process.

Assessments, Performance and Governance

'The most vocal opponents of assessment know deep down in their hearts that they are mediocre teachers.'
Presenter commentary at assessment conference, June 2013

In this final section I take up Watts' third vertex, namely, the question of how changes to an existing order affect everyday lives in ways that prompts cultural strategies – often contradictory – aimed at carving out more satisfactory alternatives. This particular aspect of Watts' analytical approach drew in part from a remarkable series of intellectual collaborations with Allan Pred, Gillian Hart and Gunnar Olsson among others: the associated conceptual tools have influenced many. Watts has applied this approach both in his writing on the profound impact of resource 'shocks' – which inspired the Movement for the Emancipation of the Niger Delta and the earlier Maitatsine's rebellion – and in his writing about more 'everyday' struggles against the extension of neoliberal workplace relations in, among other places, contract farming and donor-funded agricultural development schemes. This section applies these concepts to assessment's 'emotional labour'.

In brief, I argue that by making assessors and professors each responsible for classroom content, university assessment engenders conflict which in turn shapes the flow of value. At the University of Kentucky, for example, the assessors' job is to check syllabi for SLOs, collect representative assignments and organize 'norming' sessions in which these assignments are reviewed to see whether their content conforms to syllabi-listed SLOs. For their part, professors are expected to perform the assessed activities, design syllabi that integrate SLOs (some of which may be required by university curricular guidelines) and determine how those outcomes will be achieved and graded.

These assessment procedures apply more generally to SACSCOC accredited schools. Beyond schools in the US south, such procedures have gained ground particularly in private for-profit and non-branded public universities. Their advocates include accreditation agencies, professional groups such as the Association for the Assessment of Learning in Higher Education, and university administrators interested in 'accountability.'[7] Yet assessment has also encountered resistance. Branded institutions – the University of Kentucky president's office for instance – prefer to market their 'brand' rather than rely on comparative metrics. This points to the complexities of institutional response: branded institutions such as the University of Kentucky show a diversity of internal orientations towards assessment – assessor's interest, professor's foot-dragging, a Dean's engagement with accreditors,[8] the president's office's concern with checking accreditor power. Other schools also have asserted their autonomy vis-à-vis accreditors, as when the City University of San Francisco challenged and successfully overturned one national body's denial of accreditation.[9]

This diversity in institutional response is evident also within SACSCOC-accredited schools, which show a wide variation in both assessment procedures and professors' everyday resistance to them. Two examples taken from keynote presentations at the 2013 annual AAHLE conference suggest that the political economy of assessor/professor power relations tends to favour the assessor in non-branded institutions (particularly where adjunct teaching predominates) and the professor in branded universities. The first example describes adjunct-taught courses in a professional accreditation programme:

> [O]ur program ... is an on-line, accelerated 1-year program that serves 26 rural counties ... Outcomes ... help us keep our SACS accreditation ... it can also be close to your demise [*laughter*] ... we mapped every assessment to see ... types of students ... at risk. [T]he dismal rates that almost got us shut down went to 100% in one year. Being the only full-time employee and everyone else being adjunct, assessments have meant ... time-saving ... and we still love outcomes! (4 June 2013, AAHLE, Lexington KY)

In this case, the speaker was the professional credentialling institution's one full-time employee – an assessment manager tasked with analysing student outcomes in order to get students through a state board exam. This manager wielded significant authority over a widely dispersed adjunct teaching corps and was able to direct adjunct efforts down to the level of individual student portfolios.

This second example describes an assessor's experience of creating protocols for professor-taught classes in an R1 (Research 1 designated) university.

> [I]n our group we did have structured, meaningful conversations about assessing student learning, and there was a sort of mechanism in place to compel us to try out our ideas ... We were embroiled in what I like to call operational details ... developing forms, publishing details, forming committees ... [T]he biggest crisis point ... was when the small, highly motivated working group had to go back and talk to the faculty in their schools and in their departments ... it didn't work ... [W]e wrote policies and procedures ... and really thought that by writing the practices this meant that they would really happen. (4 June 2013, AAHLE, Lexington KY)

In this second case, the assessor had relatively little control over tenured faculty, and teaching was not typically directed towards professional training and board certifications. To make up for their limited power, assessment staff worked to create a system that would draw professors into an engagement with 'scholarly teaching'.

Assessment, in these vignettes, is shown to be a form of interactive and indeed emotional service labour, in that assessors often must cajole or entice professors to adhere to approved SLOs and rubrics, and to perform the tasks required to assess them such as turning in representative assignments. In managing the professoriate, assessors seek to shape the context of service provision, just as do other interactive service workers such as airline stewards, hairdressers and fast food counter employees (Hochschild 1983; Leidner 1993; McDowell 2008). But their emotional labour differs from others in that it runs up against a structural conflict: administrators have tasked both professors *and* assessors with responsibility for course SLOs, but only professors have direct access to curricular content in their syllabi and classrooms. Thus, unlike on airline flights or in hamburger chains, the client – in this case a professor – may have little interest in consuming the product on offer, namely assessment services.

Adjuncts may have little choice but to consume the product, due to assessors' authority to request their syllabi and student dossiers, and to direct how courses are taught. In the R1 university context, by contrast, assessors typically lack this authority, and therefore must often work much harder to find and hold on to consumers of their services.[10] The second vignette illustrates the type of difficulties they encounter. That assessor's engagement with a 'small, highly motivated' faculty group languished, finally reaching a 'crisis point' when it became clear that this effort had failed to ignite broader faculty interest in assessment.

In sum, assessors and professors work in parallel. Both are expected to focus on course content as a path to improving student outcomes, yet in doing so they confront very different political economies of value. As college instructors, our performances and even our own bodies become a part of the commodity, particularly when course evaluations may figure directly in salary calculations. We tend to view the syllabus and content delivery as an extension of our identities (McDowell 2008; Leidner 1993; Benjamin 1968). Assessors' positionality is obviously very different, since their objective is the articulation of SLOs across classrooms where their work may or may not be welcome.

This rather pedestrian case of SLOs highlights how struggles over labour's value are sometimes waged through struggle over protocols that themselves shape meaning. Assessors' labour in particular serves to specify the practices that determine which qualities become knowable and how they are qualified. Practices such as creating rubrics, collecting artifacts and norming show how the power to set protocols gives assessors a potentially outsized influence over the meaning of assessed SLOs. When these SLOs include items such as 'citizenship' – as in the case of common core curriculum – the manner in which related learning objectives are specified may have far-reaching consequences. Thus, although assessment efforts remain contested, they have already served to reposition professors and assessors.

Conclusion

Certification and accreditation provide one form of everywhere to privatization's everything, providing 'pipes and tubes' through which value – equated across space and stripped of risk – may be stabilized and realized. By applying Watts' tripartite analytical schema to education, I have identified three dimensions of governance-by-assessment. First, examining the political economy of credit transfers shows how it permits an education available in more expensive, 'destination' institutions to be subsidized by contingent faculty in non-branded institutions. It also shows how the creation of the legal and institutional framework necessary for credit transfers has received support from student loan financiers and accreditors that stand to benefit from them. Second, analysing standardization discourses reveals how the ideal of a standards-based 'common language' serves to mask the structural antinomies and inequalities underlying value transfer, just as the evocation of 'risk' serves to 'naturalize' the social exploitation on which value is written. Finally, examining assessment's emotional labour reveals how this exploitation may be further stabilized if a particular type of neoliberal subject finds satisfaction in successful assessment. Altogether, this attention to political economy, discourse and everyday work shows how educational content that is 'certified everywhere' is built on the casualization of labour, and in support of the 'privatization of everything'.

Notes

1   Regional educational accreditors do not appear to conform to ISO norms. According to ISO, 'Accreditation is the procedure by which an authoritative body gives formal recognition that a body or person is competent to carry out specific tasks …' (http://www.ansi.org/accreditation/faqs.aspx). However, inspections are typically expected to be performed by third party inspectors that are substantially independent, a norm not rigorously applied by accreditors.

2   http://ecs.force.com/mbdata/mbquestRT?rep=TA02 (accessed 9 April 2016): the Education Commission of the States maintains a multi-state comparison database of transfer policies and construction equivalencies that become routinized as automated transfer links.

3   http://www.learningoutcomesassessment.org, an organization funded by Lumina foundation: Lumina is linked to student loan financer Salle Mae (see Daden 2014).

4   Per-state fees detailed on the College Board online-only table: https://trends.collegeboard.org/college-pricing/figures-tables/tuition-fees-sector-state-over-time (accessed 22 July 2016).

5   https://www.aaup.org/sites/default/files/Faculty_Trends_0.pdf (accessed 15 May 2017); Edmonds (2015).
6   Image: https://www.youtube.com/watch?v=TAk335NOOoU (accessed 16 April 2016).
7   In 2013 I attended the annual AALHE conference in Lexington Kentucky.
8   The (previous) Dean's embrace of assessment points to the complex ways in which curricular assessment plays out in academe. He was much taken by Derek Bok's *Our Underachieving Colleges* – which he purchased by the box and distributed widely, posing that as a rationale for adopting a common core curriculum. It also provided a resume-building exercise, which likely contributed to his securing of a Provost's position in another university.
9   http://www.aft.org/node/10131 (accessed 15 May 2017).
10 An instance of indirect assessment is found in the example of SLO 'norming'. In this process, assignments are collected from each course sharing a particular SLO rubric. These assignments are read and 'normed', a practice in which each assignment is assigned a numerical rating in accordance with the degree to which it conforms to the respective SLO. Assessment reports are sent to department chairs who then pass this information on to professors who teach courses with a given SLO.

References

American Federation of Teachers. 2010. National survey of part-time/adjunct faculty, *American Academic*, vol. 2, pp. 1–15. http://www.aft.org/sites/default/files/aa_partimefaculty0310.pdf (accessed 29 July 2016).

Andreotti, V. 2006. Soft versus critical global citizenship education, *Policy and Practice: A Development Education Review*, vol. 3, pp. 40–51.

Aud, S., Fox, M. A. and KewalRamani, A. 2010. *Status and Trends in the Education of Racial and Ethnic Groups*, National Center for Education Statistics, Education Statistics Services Institute – American Institutes for Research NCES 2010-015 US Department of Education.

Ball, S. J. 2012. Performativity, commodification and commitment: an I-spy guide to the neoliberal university, *British Journal of Educational Studies*, vol. 60, no. 1, pp. 17–28.

Benjamin, W. 1968. Paris, capital of the 19th century, *New Left Review*, vol. I, no. 48, pp. 77–88.

Bok, D. 2006. *Our Underachieving Colleges: A Candid Look at How Much Students Learn and Why They Should Be Learning More*. Princeton, NJ, Princeton University Press.

College Board. 2015. *Trends in College Pricing*, College Board. http://trends.collegeboard.org/sites/default/files/2015-trends-college-pricing-final-508.pdf (accessed 28 July 2016).

Daden, D. 2014. Elizabeth Warren faces right-wing stooge: here's who's quietly funding her top critic, *Salon*, 11 June 2014. http://www.salon.com/2014/06/11/elizabeth_warren_faces_right_wing_stooge_heres_whos_quietly_funding_her_top_critic/ (accessed 28 July 2016).

Davies, B. and Bansel, P. 2010. Governmentality and academic work: shaping the hearts and minds of academic workers, *Journal of Curriculum Theorizing*, vol. 26, no. 3, pp. 5–20.

Dolan, C. S. 2008. In the mists of development: fairtrade in Kenyan tea fields, *Globalizations*, vol. 5, no. 2, pp. 305–318.

Edmonds, D. 2015. More than half of college faculty are adjuncts: should you care?' *Forbes*, 28 May 2015. http://www.forbes.com/sites/noodleeducation/2015/05/28/more-than-half-of-college-faculty-are-adjuncts-should-you-care/#4b68a4cd1d9b (accessed 28 July 2016).

Harney, S. and Moten, F. 2013, *Undercommons: Fugitive planning and black study*. New York, Autonomedia.

Hochschild, A. R. 1983. *The Managed Heart: Commercialization of human feeling*. Berkeley, University of California Press.

Hudson, I. and Hudson, M. 2003. Removing the veil? Commodity fetishism, fair trade, and the environment, *Organization & Environment*, vol. 16, no. 4, pp. 413–430.

Leidner, R. 1993. *Fast Food, Fast Talk: Service work and the routinization of everyday life*. Berkeley, University of California Press.

McDowell, L. 2008. *Working Bodies: Interactive service employment and workplace and workplace identities*, Malden, MA, Wiley Blackwell.

Mitchell, K. 2003. Educating the national citizen in neoliberal times: from the multicultural self to the strategic cosmopolitan, *Transactions of the Institute of British Geographers*, vol. 28, no. 4, pp. 387–403.

Mott, C., Zupan, S. Debbane, A.-M. and L., R. 2015. Making space for critical pedagogy in the neoliberal university: struggles and possibilities, *ACME: ejournal of critical geography*, vol. 14, no. 4, pp. 1260–1282.

Mountz, A., Bonds, A., Mansfield, B., Loyd, J., Hyndman, J., Walton-Roberts. ... Curran, W. 2015. For slow scholarship: a feminist politics of resistance through collective action in the neoliberal university, *ACME: An International E-Journal for Critical Geographies*, vol. 14, no. 4, pp. 1235–1259. Mutersbaugh, T. 2005. Just-in-space: certified rural products, labor of quality, and regulatory spaces, *Journal of Rural Studies*, vol. 21, no. 4, pp. 389–402.

Otto, J. and Mutersbaugh, T. 2015. Certified political ecology, in T. Perreault, G. Bridge and J. McCarthy (eds), *The Routledge Handbook of Political Ecology*. New York, Routledge, pp. 418–430.

Slaughter, S. and Rhoades, G. 2000. The neo-liberal university, *New Labor Forum*, vol. 6, pp. 73–79.

Spivak, G. C. 1993. *Outside in the Teaching Machine*. New York, Routledge.

Walker, J. 2009. Time as the fourth dimension in the globalization of higher education, *The Journal of Higher Education*, vol. 80, no. 5, pp. 483–509.

Watts, M. J. 1993. Development I: power, knowledge, discursive practice, *Progress in Human Geography*, vol. 17, no 2, pp. 257–272.

Watts, M. J. 1994. Development II: the privatization of everything?, *Progress in Human Geography*, vol. 18, no. 3, pp. 371–384.

Watts, M. J. 2001. Black acts, *New Left Review*, vol. 9, pp. 125–139.

Watts, M. J. 2012. A tale of two gulfs: life, death and dispossession along two oil frontiers, *American Quarterly*, vol. 64, no. 3, pp. 437–467.

10

Oil, Indigeneity and Dispossession

Joe Bryan

The headline of the press release commanded attention before the fax transmission was complete: 'Grief for the assassination of Ken Saro-Wiwa and Eight Ecologists.' None of us working in the offices of the Amazanga Institute for Indigenous Science and Technology in Puyo, Ecuador knew much about Saro-Wiwa or the Nigerian state that had just executed him and eight of his colleagues. Not that it mattered. Their deaths spoke to the threat of oil violence that dominated the political landscape in which Amazanga was immersed. Since the 1930s, the surrounding area had been shaped by oil speculation. Royal Dutch Shell, the company directly involved in Saro-Wiwa's death, was the first to survey the region's oil potential, relying on missionaries and the Ecuadorian army to protect its crews from attacks by indigenous residents (Muratorio 1991). From 1965 to 1993, United States-based Texaco developed an oil field north of Puyo, devastating a vast area of Amazonian forest, leaving Kichwa, Cofán and Secoya people with toxic levels of pollution in a largely denuded landscape similar to conditions described in the Niger Delta (Sawyer 2004; Valdivia 2005). Amazanga had been founded to prevent a similar catastrophe in Pastaza province, launched by the Organization of Indigenous Peoples of Pastaza to document indigenous use of forest resources and craft an alternative future free from oil development. That context informed Amazanga's swift reply to the news from Nigeria. Speaking from the throes of their own fight against oil companies, Amazanga's reply reflected its 'profound grief over the deaths of the nine comrades (*compañeros*) in Nigeria, affirming our solidarity with the Ogoni people because the indigenous struggle is

Other Geographies: The Influences Of Michael Watts, First Edition. Edited by Sharad Chari, Susanne Freidberg, Vinay Gidwani, Jesse Ribot and Wendy Wolford.
© 2017 John Wiley & Sons Ltd. Published 2017 by John Wiley & Sons Ltd.

a global one, and it demands that we maintain solidarity with the great
diversity of ways in which Indigenous Peoples have fought against colonialism'. It was 14 November 1995.

Oil's capacity as a destroyer of worlds is as well-known as its reputation for generating fabulous profits and corruption. These qualities, as
Michael Watts has argued for some time, are not discrete problems that
can be solved in isolation through better governance, corporate accountability or weaning modern capitalism off its dependency on oil (Retort
2005; Watts 2007). Rather they must be thought together as parts of an
'oil assemblage' whose totalizing force conjoins oil companies, state
officials, security apparatuses and local communities (Watts 2012). This
assemblage not only organizes the relations of production that produce
oil as a commodity. They also shape the very conditions of life itself, as
Amazanga's statement makes clear. Watts' concern is with how the pollution, violence and destruction inflicted by oil dispossesses communities while drastically limiting their access to the new one created by oil
wealth. In the shadow of the accumulation of tremendous wealth from
oil, Delta communities live in one of the more polluted environments in
the world, with diminished means for securing food, water and shelter.
These are the conditions that Amazanga recognized as similar to their
own in their declaration of solidarity with the Ogoni, a resemblance
made visible by oil yet understood in terms of indigeneity. Thinking
with Amazanga's statement demonstrates the capacity of indigeneity as
a descriptive, analytical category, a means of linking people and places
as different from each other as the Niger Delta is from the Amazonian
lowlands of the Ecuadorian Oriente. Mobilized in response to dispossession, indigeneity is much more than a reaction to the integrated
forces of capitalism and colonialism. It provides a means of challenging
the constructions of subjectivity and geography that underwrite analysis,
refusing to dissolve differences into a linear, universalizing narrative.
Instead, indigeneity provides a means of thinking 'contrapuntally',
holding the differences in concert with one another in ways that are
generative of new categories of thought and action (Said 1993; see also
Castree, Featherstone and Herod 2008). The contrapuntal geographies
in Amazanga's statement help us reconsider the forms that opposition
to extractivism might take.

Extractivism and the Southern Question

One of the signature effects of the oil assemblage is its transformation of
vast areas according to the needs of oil production. As Uruguayan journalist Raúl Zibechi observes, this 'scorched earth model' of commodity

production 'creates a society without subjects' in which the dispossessed are reduced to mere objects (Benito 2015). Zibechi's line revisits Marxist debates over the outcome of primitive accumulation as a regime of wealth accumulation that operates through dispossessing people of their means of production, namely land, compelling them to sell their labour through entry into the proletariat. Marx used the concept to consider the importance of colonialism to capitalism, identifying capitalism's structural dependency on spatial expansion. Oil production illustrates this point, reliant on the creation of a 'permanent frontier' (Watts 2012). As Watts makes clear, the frontier does not consist of an area outside capitalism. Rather it is a space produced by capitalism through the identification of a lack of economy that is further used as cause for intervention (Mitchell 2007). Unlike older forms of agriculture or even colonialism, the dispossession inflicted by oil production does not create an immediate means for including the dispossessed through labour markets. Much more than a means of accumulation, dispossession becomes a terminal condition that removes social lines of mobility and complicates the prospects for recognizable political action. Unable to become political subjects, the dispossessed are reduced to objects to be managed or controlled in order to minimize any threat to oil production. Zibechi uses this impasse to highlight a problem with 'extractivism' in general, sacrificing large areas to the extraction of natural resources such as oil, minerals or even industrial agriculture to export-oriented production (Acosta 2013; Gudynas 2009; Svampa 2013).

In much of the debate on extractivism in Latin America, the overarching concern has been with defining it as a mode of economy that can be further used to locate 'national' economies within global capitalism (Veltmeyer and Petras 2014). As insightful as this project has been, it often lacks attention to the historically and geographically contingent aspects of extractivism foregrounded by Watts' 'oil assemblage'. In particular, it complicates analysis of dispossession as a condition. Where discussions of extractivism take up the question of political subjectivity, analysis is often located with the terrain of the nation and citizenship (Perreault and Valdivia 2010; Valdivia 2005). The 'indigeneity question', to use another of Watts' (2009) phrases, tends to emphasize the recognition of cultural differences as the basis for political subjectivity. Difference provides an imperfect structure for recognizing the political subjectivity of the dispossessed, acknowledging the political economic forces that have produced their inequality without transforming their position in society. As a result, 'it is the figure of the indigenous movement – the Ogoni struggle, the *Ejército Zapatista de Liberación Nacional* (EZLN), [Kichwa] confederations, the Mayan struggle – rather than the class and accumulation question that now dominate the academic landscape'

(Watts 2009, 275). That disavowal of political economy represents a failure of analysis as much as a challenge to the movements he names.

Watts' assessment of the indigeneity question makes explicit reference to Gramsci's (2015 (1926)) much-read essay on 'The Southern Question'. The essay is often read with an emphasis on political subjectivity through a parsing of Gramsci's analysis of the failure of workers' parties in the Italian north to align themselves with the peasants in the agrarian south. In the north, party intellectuals mistakenly read the political interests of the peasants off of their structural position, assuming a 'magic formula' for winning peasant support through the redistribution of land controlled by feudal estates. The peasants lacked the ability to express anything more than their sorrow at the loss of the old feudal order, leading them into alliance with the fascists mediated by the Catholic Church. The workers' parties' interpretation of the peasants' actions as a cultural concern made matters worse, preventing them from understanding where the peasants were coming from, both literally and figuratively. Their mistake rendered the peasants subaltern, narrowing lines of social mobility. Much like the Hindu under British rule noted by Marx (1853), 'the loss of [their] old world, with no gain of a new one, imparts a kind of melancholy' in which a sense of loss overwhelms capacity for political action. The resulting problem was as much one of indeterminate political subjectivity as of the uneven geographies that separated the proletariat from the peasantry. Watts' (2003, 2004) analysis of indigeneity foregrounds the geographical aspects of this problem, singling out the failure of Saro-Wiwa's Ogoni movement to conceive of its position outside of the oil assemblage. Critics have accused him of reducing indigenous mobilization to struggles over resources, stripping them of their cultural content (Coombes, Johnson and Howitt 2012). There is something to be said for that approach, particularly with regard to how culture knits together knowledge of a place with forms of action as the basis for opposing dispossession (Coulthard 2014). What remains is the challenge of thinking indigeneity as more than an amendment to capitalism, expanding the horizon of geography as a practice through thinking with movements like Saro-Wiwa's MOSOP.

Amazanga's text provides a starting point for that task. Situating its reply firmly within the history of oil production in Ecuador, it uses indigeneity to horizontally link their struggle to the Niger Delta.

> Although we have no personal knowledge of the Ogoni People's experiences with the [oil] companies, they surely resemble ours here in the Oriente. The Companies always have the same strategy of dividing our people with the goal of exploiting and destroying the environment that sustains our way of life, as well as that of thousands of species of animals, plants, birds, and fish.

Amazanga's struggle is informed by their relationships to the world around them, highlighting the uniquely spatial qualities of their oppression (Kulchyski 2005). At the same time, those relationships are not confined to expressions of the oil assemblage. To the contrary, they provide a means of transcending it and conceiving of the possibility for non-oppressive relations (Coulthard 2014; Simpson 2011).

Spacing Indigeneity from Ogoniland to Sumak Kawsay

Amazanga's response to the execution of the Ogoni nine offers examples of how indigeneity can be used to think spatially. Indeed, Amazinga's lack of detailed knowledge of MOSOP did not hinder their recognition of the Nigerian group's efforts to carve out a space in opposition to oil development. If nothing else came of Amazinga's declaration of solidarity, it illuminated a dimension of their efforts to defend their Amazonian home from a fate similar to that of the Niger Delta. Their response did not emerge outside of capitalism. But it did not exist entirely within it either. There was no authentically indigenous place from which to begin their critique, any more than their effort could be located with the totalizing space of capital. What matters instead is the interplay of approaches, perceiving their juxtaposition in terms of finely articulated differences and grappling with their mutual constitution (Cusicanqui 2012).

On this point, Ken Saro-Wiwa's work is instructive. At the height of his leadership, Ogoniland 'was [a] minority backwater which by the 1990s accounted for less than 1.2 per cent of Nigerian oil' (Watts 1997, 36). Within Ogoniland, things looked very different. The threat posed by oil to life and livelihood was comprehensive, a situation that Saro-Wiwa (1992) described as 'genocide'. That charge provided a means of mobilizing an Ogoni demand for autonomy that was 'at odds with both the ideology of a Nigerian federation and with northern hegemony over a fragile and contested polity without a robust sense of national identity' (Watts 1997, 37). Ogoni demands offered an antipode to the state's claim to represent the nation. Ogoniland was more than an object of political contest. It was an idea that informed both a critique of the Nigerian state and oil production, grounded in a spatial understanding that foregrounded the reciprocal relations and obligations constitutive of a collective way of life.

Saro-Wiwa's grasp of the spatial aspects of this struggle catapulted MOSOP to international attention. Where previous efforts to draw international attention to the genocide of the Ogoni people had failed to generate international attention, connecting their minority struggle with environmental rights proved a potent mix (Nixon 2011). Tellingly, Saro-Wiwa's use of indigeneity came about through his international work in the early 1990s that helped position the Ogoni struggle within a constellation of people and places. Writing from a prison in Sani Abacha's Nigeria in 1994, Saro-Wiwa (1995, p. 130) commented:

> Contrary to the belief that there are no indigenous people in black Africa, our research has shown that the fate of such groups as the Zangon, Kataf, and Ogoni in Nigeria are, in essence, no different from those of the Aborigines of Australia, the Maori of New Zealand and the Indians of North and South America. Their common history is of the usurpation of their land and resources, the destruction of their culture and the eventual decimation of the people.

What words like that meant in Ogoniland was another matter. As Saro-Wiwa was well aware, there was no unifying notion of cultural identity that alone could provide the organization and discipline needed to fend off Shell and the Abacha government (Watts 1997). There were other ethnic groups more directly affected by oil production with equally valid claims against the state and oil companies (Okonta and Douglas 2001; Watts 2010). Against these tensions, Ogoniland was its own imaginative geography, a means for understanding the lethal combination of state force and petro-capitalism in internationalist terms. In retrospect, Saro-Wiwa's execution may well have been the high point of that imaginary. Though his work went on to inspire movements across the Niger Delta, divisions in the region itself soon eclipsed its visibility. Corrupt chiefs used culture to shore up their gerontocratic authority and exploit differences within MOSOP, while a new generation took up arms in an attempt to ransom oil production. As the spokesperson for the most visible of these groups, the Movement for the Emancipation of the Niger Delta (MEND), proclaimed, 'we are not communists or even revolutionaries. We're just extremely bitter men' (Watts 2010, 38). Statements like that confirmed the worst of Saro-Wiwa's fear that after his death 'conditions across the oilfields [would remain] the same, only worse' (Watts 2010, 37).

Oil extraction in the Amazonian lowlands of Ecuador offers a point of comparison. Upon learning of the execution of the Ogoni nine, Amazanga had this to say:

> The struggle to defend our territories and affirm our human right to self-determination and to make our own future is a struggle that we share with all indigenous peoples of the planet. We will fight to the end to defend our rights as a people, a struggle that is inextricably linked to care for the environment.

They wrote from experience. In 1992, OPIP led a march of 2000 indigenous peoples from their home in lowland Amazon to the Ecuadorian capital in Quito to protest oil development in Pastaza. The Ecuadorian state responded with a pledge to title indigenous lands. Upon returning to their lowland communities, OPIP and its constituents learned that the state had granted a new oil concession in the middle of these same lands to the United States-based company ARCO. The Ecuadorian government's rationales for the concession hinged on their insistence that subsurface resources remained the sole property of the state and were to be used for the benefit of the nation (Sawyer 2004).

Amazanga was founded by OPIP as a means to counter the concession, not by facing off directly with the oil companies but instead through a careful and close study of how OPIP's primarily Kichwa constituents made their lives living in the forest. That vision drew from a 1991 study commissioned by OPIP (Viteri, Tapia, Vargas, Flores and González 1992). Directed by one of OPIP's founders, the study presented an inventory of the 'forms of natural resource management in indigenous territories of Pastaza province, Ecuador'. Its methods were scientific, substantiating indigenous practices through soil analysis, classification of ecosystems and surveys. The analysis, however, was unmistakably Kichwa. Against the history of missionization, rubber tapping and oil extraction, the document proposed the importance of reinvigorating a collective way of life referred to with the Kichwa phrase 'sacha runa yachay' – the knowledge/way of life (yachay) of forest (sacha) people (runa). Three concepts sharpened their analysis: sumak allpa, sacha kawsay riksina and sumak kawsay. Taken consecutively, they referred to maintaining the equilibrium between people and environment that approximates the idea of territory; the art of acquiring, sharing and using knowledge derived from one's surroundings; and the 'theory and practice that shows how to live'. This last point, sumak kawsay, was further defined as the 'unifying substrate of everything in life' that threaded together the various worlds (pacha) that structure life and knowledge. Written up after the 1992 march, the document elaborated on the rallying cry of the 1992 march: 'allpamanda, kawsaymanda, ¡jatarishun!' – for earth, for life, we rise up!

By the time of Saro-Wiwa's execution, the unity of the march was passing into memory. In the concession area granted to ARCO, evangelical missionaries had created a new indigenous organization independent of OPIP. This new group had negotiated with ARCO to define the terms of entry, allowing for drilling of a well in exchange for building a schoolhouse, a community meetinghouse and a medical dispensary (Sawyer 2004, 68). The agreement modelled practices of securing informed consent, though the legitimacy of its indigenous party was disputed.

Elsewhere, OPIP affiliated communities were clamouring for alternatives that would deliver the benefits associated with oil to them – schools, health care and royalty payments. Rumours of alternative indigenous organizations negotiated separate deals with oil companies were rampant. On top of it all, the Ecuadorian government was poised to put another round of oil concessions up for bid.

It was from within this context – this space – that Amazanga responded to the news of Saro-Wiwa's death, recognizing the juxtaposition of points in common – state violence, petrocapitalism – with points of difference. Following Watts, that juxtaposition is evocative of complex interactions and ties that dispel any idea of defending the local against external forces. Much like MOSOP under Saro-Wiwa, Amazanga's position was fashioned from the weaving together of these elements, of Western science and indigenous knowledge, global capital and indigeneity. Amazanga's efforts did not seek to reduce the differences of these elements to a single logic. Instead they put their differences into play with one another, allowing them to antagonize and complement each other in ways conducive towards expanding the terrain of their opposition to oil development. Rather than struggles for land and resources, Ogoniland and *sumak kawsay* demonstrate an understanding of territory as a set of relations and obligations that could productively be used to make sense of the destruction visited by capitalism on peoples' lives and worlds. This is what differentiates their use of territory from struggles over land as a means of production. The difference is crucial to understanding their ability to provide a normative basis for critiquing capital (Coulthard 2014).

The difference between territory and land is often ignored. In Ecuador and Latin America more generally, indigenous demands for territory and autonomy have increasingly been recognized in terms of communal rights to property that obscure tensions that shape demands like those made by the Kichwa communities. In 2012, the Correa administration made good on the 1992 promises of land titling by demarcating and formally recognizing the Kichwa Nation's ownership of the territory used and occupied by its members. The Ecuadorian state's actions followed a broader trend in Latin America towards recognition of communal rights to property (Bryan 2012). Tellingly, their recognition of land rights was followed almost immediately by putting up for international bid the rights to the oil underneath Kichwa lands. The predicament is made all the more acute by the fact that *sumak kawsay* – better known now by its Spanish translation, *buen vivir* – has been formally incorporated into the Ecuadorian constitution in 2008, along with reforms recognizing the state as pluri-national and multi-ethnic (Acosta 2008; Gudynas 2010). The concept's potential to break with neoliberal policies has been undermined by the Correa administration's commitment to extractivism as the basis for the national economy (Perreault and Valdivia 2010; Radcliffe 2012). The Correa administration's move is emblematic of the 'new extractivism' in Latin America (Acosta 2013; Gudynas 2009; Veltmeyer and Petras 2014). Over and against its proclaimed opposition to neoliberalism, it affirms the fundamental importance of extractivism in the growth of the national economy.

A similar scenario has played in the administration of Goodluck Jonathan in Nigeria. Born and raised in the Niger Delta, Jonathan's cabinet included a number of anti-oil activists. Jonathan implemented a tenuous ceasefire with MEND, a move that created the stability so desired by the oil companies while 'creating lucrative opportunities for the political class and regional elites in and outside of government' through new vehicles for distributing oil rents and revenues (Watts 2015, 9). It is worth noting that all of these models, regardless of whether they claim to be post-neoliberal or not, are founded on dispossession as an inevitable component of extractivism. If they include indigenous peoples, it is only provisionally through recognition of their rights to property and promises of future inclusion. Recognition of cultural differences comes at the disavowal of critiques of political economy (Watts 2009).

If indigeneity cannot fail to engage in critique of political economy, Marxists cannot fail to grasp the importance of indigeneity. Curiously, David Harvey has been among the more prominent Marxists to dismiss indigeneity for its failure to offer a viable 'template for more global anti-capitalist solutions' (Harvey 2012, 122). Harvey has followed up this sentiment by holding out the Correa administration in Ecuador and the Morales administration in Bolivia as leading examples of post-neoliberal extractivism (Carro 2014). For Harvey, the dilemma in both countries is not how they deal with indigeneity. Instead it concerns how they negotiate the contradiction between extractivism with the constitutional mandates imposed by *buen vivir* through state-led distribution of benefits (Carro 2014). Harvey's position demonstrates not only failure to seriously engage with indigeneity and the politics of social movements more generally (Morton 2013; Reyes 2015). It also fails to take extractivism seriously as a form of economy, in particular its serial dispossessions (Martínez Alier 2015). Much like the Northern intellectuals, it mistakes redistribution for a 'magic formula'.

Conclusion: Extractivism, Indigeneity and Marxism

The problem of oil, like that of extractivism, is much more than an economic one. Responding to it requires a more comprehensive approach, attending to its modes of economy, forms of power, the forms of life it makes (im)possible. Ogoniland and *sumak kawsay* represent two different efforts to that challenge. Both conceive of the problem grasped in terms of oil being more than an economic concern. The threat of dispossession cannot be adequately addressed through recognition of indigenous land rights, improved forms of state redistribution or better drilling technology. It requires confronting the totalizing force of oil assemblage. Ogoniland and *sumak kawsay* are more than objects of recognition or concepts to fill analytical gaps. They are concepts to think with and learn from. Each guides new forms of sociality fashioned to counter the destructive force of dispossession, forging new worlds rather than filling in analytical blank spots. Michael Watts' work has been tremendously important to opening up that conversation, above all through its lengthy engagement with the legacy of Ken Saro-Wiwa, reminding us that his is not the first, nor the last, drop of blood in your oil. It is that emphasis on the violence of dispossession and how it shapes political mobilization upon which not only Watts, but Saro-Wiwa and many others like him insist that we focus our attention.

References

Acosta, A. 2008. El buen vivir, una oportunidad por construir, *Ecuador Debate*, vol. 75(December), pp. 33–47.

Acosta, A. 2013. Extractivism and neoextractivism: two sides of the same curse, in M. Lang and D. Mokrani (eds), *Beyond Development: Alternative Visions from Latin America*. Quito and Amsterdam, Fundación Rosa Luxemburg and the Transnational Institute, pp. 61–86.

Benito, V. 2015. 'Extractivism creates a society without subjects': Raúl Zibechi on Latin American Social Movements. Retrieved 19 February 2016 from http://upsidedownworld.org/main/international-archives-60/5409-extractivism-creates-a-society-without-subjects-raul-zibechi-on-latin-american-social-movements

Bryan, J. 2012. Rethinking territory: social justice and neoliberalism in Latin America's territorial turn, *Geography Compass*, vol. 6, no. 4, pp. 215–226.

Carro, H. 2014. David Harvey: Latinoamérica toma la posta para 'revertir los peores aspectos del neoliberalismo', *ANDES*, retrieved 14 July 2016 from http://www.andes.info.ec/es/noticias/david-harvey-latinoamerica-toma-posta-revertir-peores-aspectos-neoliberalismo.html

Castree, N., Featherstone, D. and Herod, A. 2008. Contrapuntal geographies: the politics of organizing across sociospatial difference, in K. R. Cox, M. Low and J. Robinson (eds), *The SAGE Handbook of Political Geography*. London, SAGE Publications Ltd, pp. 305–321.

Coombes, B., Johnson, J. T. and Howitt, R. 2012. Indigenous geographies I: mere resource conflicts? The complexities in Indigenous land and environmental claims, *Progress in Human Geography*, vol. 36, no. 6, pp. 810–821.

Coulthard, G. S. 2014. *Red Skin, White Masks: Rejecting the colonial politics of recognition*. Minneapolis, MN, University of Minnesota Press.

Cusicanqui, S. R. 2012. Ch'ixinakax utxiwa: a reflection on the practices and discourses of decolonization, *South Atlantic Quarterly*, vol. 111, no. 1, pp. 95–109.

Gramsci, A. 2015 (1926). *The Southern Question*, Translated by P. Verdicchio. New York, Bordighera Press.

Gudynas, E. 2009. La ecología política del giro biocéntrico en la nueva Constitución de Ecuador, *Revista de estudios Sociales*, vol. 32, pp. 34–46.

Gudynas, E. 2010. Si eres tan progresista ¿por qué destruyes la naturaleza? Neoextractivismo, izquierda y alternativas, *Ecuador Debate*, vol. 79, pp. 61–81.

Harvey, D. 2012. *Rebel Cities: From the right to the city to the urban revolution*. London, Verso Books.

Kulchyski, P. K. 2005. *Like the Sound of a Drum: Aboriginal cultural politics in Denendeh and Nunavut*. New York, Cambridge University Press.

Martínez Alier, J. 2015. Gudynas y Harvey, *Rebelión*, retrieved 5 December 2015, from http://www.rebelion.org/noticia.php?id=204595

Marx, K. 1853. The British rule in India, *New-York Daily Tribune*, 25 June, retrieved 25 November 2015 from https://www.marxists.org/archive/marx/works/1853/06/25.htm

Mitchell, K. 2007. Geographies of identity: the intimate cosmopolitan, *Progress in Human Geography*, vol. 31, no. 5, pp. 706–720.

Morton, A. D. 2013. *Revolution and State in Modern Mexico: The political economy of uneven development*, Lanham, MD, Rowman & Littlefield.

Muratorio, B. 1991. *The Life and Times of Grandfather Alonso, Culture and History in the Upper Amazon*. New Brunswick, NJ, Rutgers University Press.

Nixon, R. 2011. *Slow Violence and the Environmentalism of the Poor*. Cambridge, MA, Harvard University Press.

Okonta, I. and Douglas, O. 2001. *Where Vultures Feast: Shell, human rights, and oil in the Niger Delta*. San Francisco, CA, Sierra Club Books.

Perreault, T. and Valdivia, G. 2010. Hydrocarbons, popular protest and national imaginaries: Ecuador and Bolivia in comparative context, *Geoforum*, vol. 41, no. 5, pp. 689–699.

Radcliffe, S. A. 2012. Development for a postneoliberal era? Sumak kawsay, living well and the limits to decolonisation in Ecuador, *Geoforum*, vol. 43, pp. 240–249.

Retort. 2005. *Afflicted Powers: Capital and spectacle in a new age of war*. London and New York, Verso.

Reyes, A. 2015. Zapatismo: other geographies circa 'the end of the world', *Environment and Planning D: Society and Space*, vol. 33, no. 3, pp. 408–424.

Said, E. 1993. *Culture and Imperialism*. New York, Vintage Books,.

Saro-Wiwa, K. 1992. *Genocide in Nigeria: The Ogoni tragedy*. London, Lagos, and Port Harcourt, Saros International Publishers.

Saro-Wiwa, K. 1995. *A Month and a Day: A detention diary*. New York, Penguin Books.

Sawyer, S. 2004. *Crude Chronicles: Indigenous politics, multinational oil, and neoliberalism in Ecuador*. Durham, NC, Duke University Press.

Simpson, A. 2011. Settlement's secret, *Cultural Anthropology*, vol. 26, no. 2, pp. 205–217.

Svampa, M. 2013. Resource extractivism and alternatives: Latin American perspectives on development, in M. Lang and D. Mokrani (eds), *Beyond Development: Alternative Visions from Latin America*, Quito and Amsterdam, Fundación Rosa Luxemburg and the Transnational Institute, pp. 117–144.

Valdivia, G. 2005. On indigeneity, change, and representation in the northeastern Ecuadorian Amazon, *Environment and Planning A*, vol. 37, no. 2, pp. 285–303.

Veltmeyer, H. and Petras, J. 2014. *The New Extractivism in Latin America: A post-neoliberal development model or imperialism of the twenty-first century?* London, Zed Books.

Viteri, A., Tapia, M., Vargas, A., Flores, E. and González, G. 1992. Plan Amazanga: Formas de manejo de los recursos naturales en los territorios indígenas de Pastaza, Ecuador, Organización de los Pueblos Indígenas de Pastaza (OPIP), Puyo, Ecuador.

Watts, M. 1997. Black gold, white heat: state violence, local resistance and the national question in Nigeria, in S. Pile and M. Keith (eds), *Geographies of resistance*, London, Routledge, pp. 33–67.

Watts, M. 2003. Development and governmentality, *Singapore Journal of Tropical Geography*, vol. 24, no. 1, pp. 6–34.

Watts, M. 2004. Antinomies of community: some thoughts on geography, resources and empire, *Transactions of the British Institute of Geographers*, vol. 29, 195–216.

Watts, M. 2007. Petro-insurgency or criminal syndicate? Conflict & violence in the Niger Delta, *Review of African Political Economy*, vol. 34, no. 114, pp. 637–660.

Watts, M. 2009. The southern question: agrarian questions of labour and capital, in A. H. Akram-Lodhi and C. Kay (eds), *Peasants and Globalization: Political economy, rural transformation, and the agrarian question*. New York, Routledge, pp. 262–287.

Watts, M. 2010. Sweet and sour, in M. Watts (ed.), *Curse of the Black Gold: 50 Years of Oil in Nigeria*. Brooklyn, NY, powerHouse Books, pp. 36–47.

Watts, M. 2012. A tale of two gulfs: life, death, and dispossession along two oil frontiers, *American Quarterly*, vol. 64, no. 3, pp. 437–467.

Watts, M. 2015. Chronicle of a future foretold: the complex legacies of Ken Saro-Wiwa, *The Extractive Industries and Society*, vol. 2, no. 4, pp. 607–844.

11

Frontiers
Remembering the Forgotten Lands

Teo Ballvé[1]

> *Along the perimeter of the Kano airport, I turned a corner to find the still burning fuselage of a wrecked plane in the middle of the road. Nigeria has always seemed to conjure up a heady mix of the fantastic (or perhaps I should say of magical realism), of a high-octane humanity, and prophesies of the coming catastrophe.*
>
> Watts, *Silent Violence* (2013)

> *Suddenly, as if a whirlwind had set down roots in the center of town, the banana company arrived, pursued by the leaf storm. The leaf storm had been stirred up, formed out of the human and material dregs of other towns, the chaff of a civil war that seemed ever more remote and unlikely. The whirlwind was implacable … In less than a year it sowed over the town the rubble of many catastrophes that had come before it, scattering its mixed cargo of rubbish in the streets.*
>
> Gabriel García Márquez, *The Leaf Storm* (1955)

Macondo, the fictional Latin American village setting for some of Gabriel García Márquez's stories, is your stereotypical frontier boomtown. The whirlwind arrival of a US banana company turns the sleepy village into a rowdy outpost of saloons, bordellos and slum encampments. But the boom abruptly goes bust when the company oversees the massacre of thousands of protestors during a workers' strike. With a wave of his hand, a company executive unleashes a torrential downpour that dissolves all memory of the slaughter. As suddenly as it had arrived, the banana company disappears and a government-declared state of

Other Geographies: The Influences Of Michael Watts, First Edition. Edited by Sharad Chari, Susanne Freidberg, Vinay Gidwani, Jesse Ribot and Wendy Wolford.

emergency puts Macondo under military rule. The rainstorm lasts for five years, leaving villagers slopping through muddy streets in a stupor of total oblivion.

The five-year deluge and the other occurrences of 'magical realism' that García Márquez became famous for are not what perplex the residents of Macondo. Instead, it's the forces of modernity and capitalism that are seen as mysterious and confusing (Beckman 2014). These fickle forces come and go, leaving inexplicable piles of rubble in their wake. Time and again, Macondo falls victim to the same 'wreckage upon wreckage' observed by Walter Benjamin's angel of history (its face turned toward the past and a violent storm caught in its wings). 'This storm irresistibly propels him into the future to which his back is turned,' wrote Benjamin, 'while the pile of debris before him grows skyward. This storm is what we call progress' (1968, 257).

Nowhere are these storms of progress – the gale-force winds of capitalist modernity – as palpable and destructive as in those spaces we call 'frontiers.' Paradoxically, however, frontiers are often imagined as places where these forces have not yet run their course. Frontiers, then, are spaces produced by both the power *and* the limits of reigning regimes of accumulation and rule. They achieve this feat of magical realism because frontiers are consummate examples of what Lefebvre (1991) meant by the social production of space: frontiers are ideological and discursive formations as much as material ones; they are, in every sense, real-and-imagined.[2]

Understood in this way, frontiers have been a loosely running thread through much of Michael Watts' work: from the silent violence of hunger in Hausaland, to the combustible oil politics of the Niger Delta, and just about everything else in between (from California communards to Boko Haram). The reason for this, I would argue, is that the agrarian question itself – one of his abiding concerns – has always been something of a frontier story. In this chapter, I analyse the frontier as a social-spatial formation that is produced by and is productive of what Watts calls 'economies of violence' (2004, 2006). In other words, my argument is that economies of violence are what give frontier zones their unique spatiality and, it must be said, their frightening flickers of magical realism.

My vehicle for analysing this social production of frontiers is northwest Colombia, a region torn asunder by decades of war and violent forms of accumulation. In what follows, I first show how the historical development of the region's economies of violence was wrapped up with and enabled by an insidious cultural politics emanating from the city of Medellín – an 'internal colonialism', in the language of Latin American dependency theory. Second, I illustrate how right-wing paramilitaries, which are themselves a product of this history, helped turn the region's economies of violence into a form of rule. These counterinsurgent militias helped make this frontier zone into a particular kind of 'governable space' (Watts 2004, 2006).[3]

For Watts, the term 'governable space' implies a deliberate irony; these spaces are far from orderly. In the Niger Delta, for instance, he shows how governmentality operates – if we can use that word – through the way Nigeria's economies of violence and its governable spaces are mutually produced and interlaced (Watts 2004, 2006). Far from the imposition of some panopticon-like lockdown, rule in the Delta looks a lot like unruliness. The utility of the term, then, at least for my purposes, is that it helps us see that the violent political disorder of frontiers is less an index of state absence or failure, than a constitutive part of everyday state-making in these spaces. As particular sorts of governable spaces, frontiers are sites where we can see all the messy cultural, political and economic violence of state formation in real time. But this same messiness means frontiers can also be sites of political experimentation and possibility – the space of the rebel, the bandit, the runaway, the commoner. Unfortunately, though, the story of almost every frontier is the chronicle of a death foretold.

Producing the Frontier

When the trumpet blared
everything on earth was prepared
and Jehovah divvied up the world
to Coca-Cola Inc., Anaconda,
Ford Motors, and other entities.
United Fruit Company Inc.
reserved the juiciest piece,
the central coast of my land,
the sweet waist of America.

Pablo Neruda, *Canto General* (1950)

Although literary critics often obsess about the 'magical' in García Márquez's stories, his fellow Colombians focus much more on the 'realism'. The running joke is that 'Gabo', as he's affectionately known, should've won a Pulitzer, not a Nobel. During my fieldwork, I heard frequent real-life echoes of his stories, like the old *campesino* who told me: 'These lands had no owners when we arrived, the streams and mountains didn't even have names.' It instantly conjured for me a line from the magisterial opening of *One Hundred Years of Solitude*: '... The world was so recent that many things lacked names, and in order to indicate them it was necessary to point' (García Márquez 1970, 1). Although myths about an untouched and unsoiled nature are, of course, as old as the Garden of Eden, the *campesino* was talking about Urabá, a region in the far north-west corner of Colombia, where I've been working since 2006.

He was telling me about his arrival to Urabá in the early 1960s just a few years before a land-rush incited by the whirlwind arrival of the infamous United Fruit Company. Disease was decimating the company's other plantations in Latin America, and its single enclave in Colombia near Gabo's hometown was still stained by the memory of the 1928 '*masacre de las bananeras*', the real-life event memorialized in the climax of his *magnum opus*. No one really knows how many people died, but one cheery cable from the US Embassy noted: 'I have the honor to report that the ... total number of strikers killed by the Colombian military authorities during the recent disturbance reached between five and six hundred; while the number of soldiers killed was one.'[4] Decades later, in 1964, United Fruit seized the safer shores of the soppy, steamy lowlands surrounding the Gulf of Urabá.

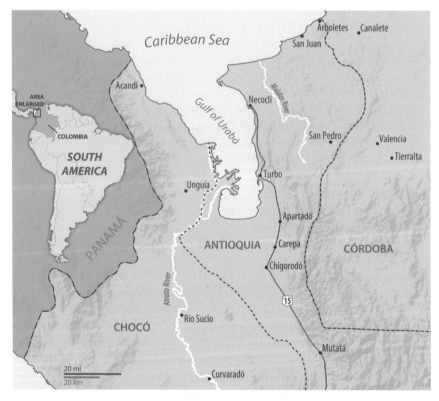

The gulf region of Urabá in northwest Colombia. Credit: Author.

The banana boom transformed Urabá at breakneck speed. UC Berkeley geographer Jim Parsons witnessed it firsthand, taking special note of Apartadó, the town at the epicentre of the Macondo-like transformation: 'It is a vast swollen slum of muddy streets and rough, palm-thatched houses without running water, or latrines. But Apartadó has three banks, a bull-ring, a radio station ('Voz de Uraba'), a newspaper (*Vanguardia de Urabá*), a modern 'subdivision,' and dozens of noisy cantinas (taverns)' (Parsons 1967, 97). The boom razed the surrounding landscape, ordering it into a geometric patchwork of lines and rectangles. As in Macondo, the real magic at work in Urabá was not some supernatural force; it was the cold, hard forces of capitalism: 'There was not much time to think about it, however, because the suspicious inhabitants of Macondo barely began to wonder what the devil was going on when the town had already become transformed into an encampment of wooden houses with zinc roofs' (García Márquez 1970, 232–233).

Frontiers are, by definition, relational spaces; they are always the frontier of somewhere, for someone. In the case of Urabá, that 'somewhere' was Medellín and the 'someone' was the city's light-skinned, famously conservative elite. By the start of Urabá's banana bonanza in the 1960s, the city had been on a roll. It was already a longstanding financial-commercial hub thanks to its historical role in the export of slave-mined gold and the import of cheap finished goods. At the beginning of the twentieth century, these merchant bankers started reaping astronomical profits by muscling their way into the burgeoning coffee boom, controlling the credit, pricing, distribution and transportation of the crop (Hylton 2007). As rising financiers, they began redirecting their excess wealth toward urban industrial development, particularly textile manufacturing.

With plenty of surplus capital on hand, they then moved on their long-standing dream of constructing the 'Highway to the Sea' connecting Medellín to the Gulf of Urabá. The city is the capital of Antioquia department and Urabá is the otherwise land-locked province's only outlet to the sea. Boosters of the Highway argued it would connect the city to maritime networks and the adjacent Panama Canal. Others dreamed of turning Urabá into cotton plantations capable of supplying the city's massive textile industry. Though cotton never took off, United Fruit stepped in at just the right moment, providing these elites with a heavily subsidized and profitable venture for mopping up their extra capital. Beyond the dull compulsions of the market, the cresting wave of primitive accumulation had a strong cultural-political current behind it.

Medellín's elites are united by their cultural identity as '*paisas*'. A play on the word '*paisano*' (countryman), '*paisa*' can transcend class positions, but at its worst it's a chauvinist regional-cultural identity articulated by people from the highlands of Antioquia (and a few neighbouring provinces). *Paisas* generally see themselves as Colombia's most enterprising, pious, conservative, hardworking, light-skinned and macho people (Appelbaum 2003). Imagine what 'Texan' means for someone like former US President George W. Bush and you have an idea of what it means to identify as '*paisa*'. Even today, it's not altogether unheard of for ultra-conservative *paisas* to proudly describe themselves as '*la raza antioqueña*' (the antioqueño race). Through this cultural prism, *paisas* saw Urabá as 'their' untapped territorial birthright – a creole version of Manifest Destiny. Urabá was also their consummate Other: a low-lying tropical zone overwhelmingly populated by Afro-Colombians and indigenous people considered incapable of taking 'proper' advantage of the region's natural-geographical endowments.

For *paisa* elites, Urabá's assumed racial degeneracy helped frame the Highway and the banana enclave as projects of moral redemption and geopolitical necessities for bringing the coastal satellite region into the orbit of Antioquia's cultural-political hegemony (Steiner 2000). They imagined *antioqueño* civilization's definitive triumph over barbarism in an area they explicitly claimed – in Imperial Roman terms – as '*mare nostrum*'.[5] For one newspaper columnist, Urabá was the unfinished business of colonial conquest: 'Let's finish what those audacious Spanish conquistadors were unable to do: subjugate and exploit that promised land' (quoted in Steiner 2000, 9).

The land speculation, dispossession and exploitation the Highway and the United Fruit Company brought in train made relations between land, labour and capital increasingly violent. Indeed, by the time communist rebels from the Popular Liberation Army (EPL) and the Revolutionary Armed Forces of Colombia (FARC) arrived to Urabá in the 1970s, they found fertile ground for their insurgencies: the region already had a strong presence of the Communist Party, an exploited and increasingly unionized banana proletariat, an impoverished peasantry, brittle government authority, well-trodden smuggling routes and plenty of jungle for guerrilla warfare. Within the broader geopolitics of the Cold War, the explosive cycle of violence between insurgency and counterinsurgency further consolidated Urabá's frontier reputation as a lawless no-man's land. By the 1980s, the EPL and the FARC had turned Urabá into a political-military bastion for the insurgencies, clinching its status as a fugitive space. But the rebels made a fateful mistake.

In the 1980s, Medellín's economy was again riding high: this time, on the cocaine boom. Pablo Escobar and his partners in the Medellín Cartel had started investing and laundering their surplus narco-capital through rural real estate and agribusiness, bringing whole new meaning to David Harvey's notion of the 'spatial *fix*' (2001) – capital gets fixed into place until, junkie that it is, it moves onto its next fix.[6] Urabá was a favourite site for these operations. For the Medellín Cartel, Urabá made economic, cultural and practical sense, giving them logistical control over a key city-to-the-sea smuggling corridor (Reyes 2009). For the region's small and medium landholders, however, the narco land-rush unleashed yet another wave of primitive accumulation. The rural oligarchy, meanwhile, responded uneasily to this ascendant class of narco estate-owners, who were grudgingly dubbed the '*clase emergente*' by their blue-blooded counterparts. But looming agrarian crises, and their intensification by neoliberalization alongside growing peasant and rural worker militancy, gave the *clase emergente* a fortuitous entrée into the upper echelons of agrarian society. However, ironically, it was the guerrillas who introduced the linchpin that sealed an alliance between the old-guard *latifundistas* and the newly minted agrarian (narco) elite.

The guerrillas made the mistake of subjecting the narco-estate owners to the same extortive kidnappings once reserved for traditional landowners. This fateful move made natural allies out of the *clase emergente* and the conservative old guard. In Raymond Williams' words, the 'overreachers' and the 'wellborn' in Colombia made common cause (1973, 61). Together, and with crucial assistance from the United States-backed military, they began forming private armies to fight off rebel aggression. In the 1990s, these right-wing paramilitary groups metastasized across the country with the drug trade as their bottomless source of funding. The counterinsurgent paramilitary movement coalesced under a loosely organized national confederation, but Urabá remained its spiritual and logistical headquarters. Although never devoid of counterinsurgent aims, the violent momentum of the growing paramilitary war machine was driven by its own internal metabolism, gaining vast amounts of lands, businesses and weapons, while eliminating political opponents and protecting its most lucrative activity, drug trafficking. Paramilitaries became the ultra-militant vanguard of Urabá's economies of violence.

Economies of Violence and Governable Spaces

> We view violence less as an epiphenomenon of economic process than a narrative and institutional force requiring its own reckoning and its own history. We view political economies as variously built in and through violence rather than, as some do, parallel forms of power, the 'economic' and the 'military.'
> Catherine Lutz and Donald Nonini (1999)

As with any social space, frontiers are made through an inseparable combination of symbolic and material processes operating at multiple scales. As I detailed above, the social production of Urabá as a frontier, for instance, hinged on racist ideologies and abstract planning as much as infrastructure, capital flows and political violence. Urabá also demonstrates how the lived experience of those living in frontier zones is typically over-determined by violent forms of accumulation. In other words, capitalism's trinity formula of land-labour-capital takes on a particularly brutal form in the production of frontiers. The social relations of land and property are particularly contested and in flux meaning that ongoing forms of primitive accumulation are woven into the very fabric of social and political struggles in such spaces. Every frontier zone has its own situated economies of violence with 'their own coherence and logic' (Watts 2012, 126), but the broad strokes of dispossession as 'ruthless terrorism' are unsparingly consistent (Marx 1990, 895).

In Urabá, paramilitaries, or '*paras*' as most Colombians call them, became the handmaidens of this violent and lawless process of agrarian counter-reform. 'From beginning to end,' explained one paramilitary commander, 'from the moment we planned, executed, and finalized our military operations, during all that time, we were buying up lands.' The peasant communities that suffered the brunt of this violence described things quite differently – nothing was actually 'bought'. As one poor farmer put it, 'They said they came here to clean out the guerrillas, but it was us, the campesinos, they cleaned out.' In interviews, several survivors said that, when the onslaught began, the *paras* came with the same bone-chilling offer: 'Sell us your land, or we'll negotiate with your widow.' As in Marx's classical account, 'All this happened without the slightest observance of legal etiquette' (Marx 1990, 884). It was a crescendo of terror that would eventually leave thousands either dead or landless. Most of this primitive accumulation occurred on the fringes of the banana enclave in places settled by *campesinos* cast out (or untouched) by previous waves of dispossession.

The *paras* turned toward a new set of agricultural ventures on the stolen lands. In the case of the Curvaradó and Jiguamiandó rivers, they chose oil palm, hoping to tap the burgeoning market in biofuels. A joint operation between the Army and paramilitaries – named 'Operation Genesis' of all things – tore through these two neighbouring basins in 1997. The *paras* and the Army collaborated so closely they usually referred to each other as '*primos*' (cousins). Forced to leave at gunpoint, many of the displaced *campesinos* took refuge in nearby towns and spent the next several years unable to visit their farms. 'All the work of my youth was gone,' recalled an elderly *campesino* about the day he first glimpsed his razed farm. Tidy and seemingly endless rows of oil palm saplings had replaced the messy patchwork of fields, pastures and forest that had previously shaped his land. The paramilitary-backed oil palm companies ended up seizing 55 000 acres of collectively titled Afro-Colombian lands in this part of Urabá alone.

They devised a viciously effective formula for executing these land grabs. After having 'cleansed' the area of the supposedly 'subversive' peasants, the *paras* repopulated small portions of the abandoned farmlands with client *campesinos* imported from elsewhere. Paramilitary operatives would then organize the incoming peasants into sham producer cooperatives (of oil palm, rubber, teak or whatever). They would parcelize the ill-gotten lands. And, finally, they would register the lands under the name of these 'community' organizations. The sequence of moves effectively laundered the illicit origins of the landholdings and gave the industrial-scale agricultural projects a congenial grassroots façade.

Through an especially perverse model of contract farming, the new occupant communities 'contributed' land and labour, while the private companies and negligent development institutions (both national and international) provided capital and technical assistance. One paramilitary-backed company alone secured the equivalent of $2.1 million from the national government using this scheme. In another case, a bloodstained teak venture – framed as a 'reforestation' project – won $250 000 dollars in grant money from the US Agency for International Development (USAID) and even more from UN agencies. USAID supported the project under the rubric of Plan Colombia – Washington's anti-drug and counterinsurgency initiative – as part of its 'alternative development' portfolio, which is aimed at weaning *campesinos* away from growing drug-related crops.[7]

According to USAID, its alternative development programmes help foster a 'culture of legality' by promoting social capital, entrepreneurship, land tenure, environmental conservation and local institution building. Within the conflated frameworks of counterinsurgency and the drug war as the subtext, USAID said the programmes 'ensure that [recipient] communities effectively transit into legality and reinforce the legitimacy of the State' (USAID 2009, 99). The national government has similarly highlighted how the projects have helped 'strengthen local institutions' while 'boosting the State's credibility and legitimating national, departmental, and municipal institutions among the communities' (Acción Social 2007, 107). In practice, these programmes put drug-war dollars into the pockets of the very drug-traffickers they claim to be fighting against. Colombians have an adjective for such unbelievable realities: *macondiano*.

The *paras*' agricultural projects could be easily interpreted as an attempt to whitewash their corporate malfeasance with fashionable and politically correct development-speak. But the problem is deeper and more serious. The grassroots development apparatus – a strategic ensemble of discourses, practices and institutional forms – became the means through which paramilitary-led primitive accumulation was executed and legalized. In the process, the *paras'* economies of violence became compatible with projects of liberal state formation associated with the imperatives of institution building, securing the rule of law, promoting good governance and attracting capital. Urabá is a testament to how Colombia's economies of violence have produced surprisingly resilient and coherent, which is not to say benevolent, formations of rule; they have been integral, in other words, to the making of a violent governable space. One paramilitary commander's boastful comment about the oil palm projects makes the link with state-building explicit:

> In Urabá, we [paramilitaries] have palm cultivations. I personally found the businessmen that invested in those projects. The idea is to take rich people to invest in those kinds of projects in different parts of the country. By taking the rich to these zones the institutions of the state also arrive. Unfortunately, the institutions of the state only back those things when the rich are there. So you have to take the rich to all those regions of the country and that's a mission shared by all the [paramilitary] commanders.[8]

The *paras* have always cited the 'absence of the state' as their main self-justification, their entire reason for being. The jailed leader of Urabá's largest paramilitary faction, a charismatic former beer truck driver known as 'El Alemán' (The German), put it plainly at the start of his trial: 'Our interest as a politico-military organization in arms was not only to win the war against Colombian society's number one enemy – the guerrillas – it was also for the state to gain a presence in those areas.' When I finally got the chance to interview him, I asked him what he meant by this. 'Look,' he began, 'the police may have had control of an area here or there without any problems, but the economic, social, political, and military power really belonged to the guerrillas.' He paused for dramatic effect and then cracked a smile: 'So what did we do? We took that power away and replaced it with our own, bit by bit.'

It was at that 'bit by bit' scale and among the select communities of *campesinos* whose support they actively cultivated that paramilitary state-building got to work. In practice, paramilitary statecraft was basically community organizing. In Gramscian terms, the *paras* began complementing their 'war of manoeuvre' against the insurgencies with a subtler 'war of position' on the politico-ideological front. Much more than a simple play for 'hearts and minds', a war of position, as Gramsci understood it, is a hegemonic struggle pressed into the service of a particular vision of statehood (Gramsci 1971, 233–239). El

Alemán trained an entire cadre from the lower-level ranks of his militia specifically for this purpose.

Christened 'Promoters of Social Development', or PDSs, his community organizers learned how to form peasant cooperatives and NGOs. They also engaged in all kinds of practical everyday affairs: from coordinating a sick person's transportation to a hospital and bankrolling Christmas parties, to the construction of basic infrastructures like roads and small bridges. Above all, however, the PDSs were especially well-versed in all the laws regulating the relationship between municipal government entities and community-scale *Juntas de Acción Comunal*. The *Juntas* are Colombia's most subsidiary institutions of local governance (think of them as a more governmentalized version of a neighbourhood association). If the production of governable space is the modality through which 'a real and material governable world is composed, terraformed, and populated' (Rose 1999, 32), then the PDSs were its midwives and the *Juntas* its institutional hubs.[9] As one former PDS told me, 'We realized we could do more by working in an organized way through the law and what was legal, than we could do with 10,000 armed men.'

Today, El Alemán, having served an abbreviated jail sentence, is continuing this work through a Medellín-based NGO he founded with some of his former lieutenants. The NGO is named (wait for it …) 'The Pro-Resilience Foundation'. The organization has a series of projects it's shopping around in the context of Colombia's emergent post-conflict transition. One of its projects proposes to help Afro-Colombian communities displaced from the Atrato River basin; the project euphemistically describes them as the 'diaspora of the Atrato'. Another proposal claims the organization will train Urabá's community leaders into what it calls 'Promoters of Resilience and Reconciliation'. More than mere window dressing for attracting money from the coming post-conflict piñata, these moves might point toward a broader shift brought on by the government's new peace deal with the insurgency: extraction and accumulation is ramping up alongside more biopolitical sorts of interventions in areas that were once off-limits. For instance, I have a photograph from Urabá that shows a well-armed soldier patrolling a brand new hospital that is located in, of all places, a free trade zone. Make live and let die; but, above all, make money.

Conclusion: Macondo Meets *Mad Max*?

Frontiers are deregulated because they arise in the interstitial spaces made by collaborations between legitimate and illegitimate partners: armies and bandits; gangsters and corporations; builders and despoilers.

Anna Tsing (2005)

As a place of 'parcelized sovereignties' (Anderson 1974; Hylton 2006), Urabá fits a standard definition of frontiers as spaces 'where no one has an enduring monopoly on violence,' as boundaries 'beyond the sphere of the routine action of centrally located violence-producing enterprises' (Baretta and Markoff 2006, 35–36). But a focus on monopolies of violence can easily gloss over the actual persistence of state structures and practices. As I have argued, Urabá's economies of violence and its governable spaces must be understood not as indices of state absence or political collapse as much as everyday forms of frontier state *formation*. Nonetheless, the geopolitical imaginary that conjures frontier zones as stateless spaces remains a powerfully productive political discourse. Frontiers, in this regard, are geographical renderings of the 'state effect' (Mitchell 1991). They produce the state as a spatial-structural effect by casting out a purportedly non-state space in desperate need of incorporation.

The cultural politics of the 'frontier effect', which is to say how the insides and outsides of state space are socially produced, are typically structured along racialized lines of colonial difference. The story of Urabá and the metropolitan gaze of *paisa* elites recall the Manichean bifurcation of space that Fanon (2004) identified as a crucial colonial practice (civilization 'over here', barbarism 'over there'). The upshot in biopolitical terms is that the agents of dispossession see the land and resources of frontier zones as utterly vital to their own survival, while subject populations are seen as utterly disposable. This internal colonialism of frontier zones is also what conditions and makes possible the near permanent suspension of the legal order. The lived experience of this putative outside is characterized by a condition where, in Benjamin's words, the 'state of emergency' is 'not the exception but the rule' (1968, 257). Macondo meets *Mad Max*. The permanence of the exception, however, stems not from frontiers existing completely outside the law, but rather because they are zones where the 'insides' and 'outsides' of the juridical order are utterly blurred and indecipherable (Agamben 2005, 23).

In Colombia, these outlaw frontier spaces are often described as the 'forgotten regions'. Hailing from one of these places, García Márquez turned forgetting and remembering into a dialectical force in *One Hundred Years of Solitude*. Following his lead, we could say that critical accounts of frontier zones – of the Macondos of our world – are also acts of remembering, a way of brushing our forgetful human geographies against the grain. Geographical inquiry should be an abrasive act: it should scrape far beneath the surface, exposing all the discomforting histories that have been buried over, silenced and ignored.

Rather than leaving frontiers to the oblivion of modernity's amnesia-inducing rainstorms, we must expose over and over how these supposedly *peripheral* spaces have in fact been *central* to the creation and maintenance of our current distribution of life in the world. Watts takes a similar stance in his new introduction to *Silent Violence*: 'Remembering surely stands at the center of what a critical intellectualism must strive for' (2013, lxxxvi). When Gabo accepted his Nobel, he also described the recollection of suppressed histories as a form of engaged praxis. Remembering, he said, is the first step for ensuring that 'the races condemned to one hundred years of solitude will have, at last and forever, a second opportunity on earth'.

Notes

1   Teo Ballvé is an Assistant Professor of the Peace & Conflict Studies Program and the Department of Geography at Colgate University.

2   Even the cursory list I had compiled about research on frontiers would have sent me well beyond my allotted (inclusive) word-count, so I limit myself to just a few citations that have been particularly influential for my thinking (Hecht and Cockburn 1989; Alonso 1995; Pratt 2008; Redclift 2006; Tsing 2005). Watts, too, has engaged directly with the concept of frontiers (1992, 2012).

3   In Watts' work from the Niger Delta, economies of violence and governable spaces are distinct yet inseparable dynamics of how the oil complex and the petro-state interconnect.

4   Cable to the Secretary of State from Jefferson Caffery, Legation of the United States of America, Bogotá, 8 December 1928.

5   *Album de la carretera al mar*, Fray Máximo de San José (Ed.), 1930, p. 90.

6   In one essay, Harvey (2001) compares capital's ceaseless need for expansion and movement to a drug addict's perpetually dissatisfied cravings.

7   I discuss the way the *paras* put 'grassroots development' to work in more depth elsewhere (Ballvé 2013).

8   'Habla Vicente Castaño', *Semana*, 5 June 2005.

9   Elsewhere, I describe how the PDSs became the midwives of Colombia's decentralization reforms at the local scale (Ballvé 2012).

References

Acción Social. 2007. *Sembramos y ahora recogemos: Somos Familias Guardabosques*. Bogotá: Acción Social & UNDOC.

Agamben, G. 2005. *State of Exception*. Chicago, University of Chicago Press.

Alonso, A. M. 1995. *Thread of Blood: Colonialism, revolution, and gender on Mexico's northern frontier*. Tucson, University of Arizona Press.

Anderson, P. 1974. *Lineages of the Absolutist State*. London, New Left Books.

Appelbaum, N. P. 2003. *Muddied Waters: Race, region, and local history in Colombia, 1846–1948*. Durham, NC, Duke University Press.

Ballvé, T. 2012. Everyday state formation: territory, decentralization, and the Narco land-grab in Colombia, *Environment and Planning D: Society and Space*, vol. 30, no. 4, pp. 603–622.

Ballvé, T. 2013. Grassroots masquerades: development, paramilitaries, and land laundering in Colombia, *Geoforum*, vol. 50, pp. 62–75.

Baretta, S. D. and Markoff, J. 2006. Civilization and barbarism: cattle frontiers in Latin America, in F. Coronil and J. Skurski (eds), *States of Violence*. Ann Arbor, University of Michigan Press, pp. 33–74.

Beckman, E. 2014. What's left of Macondo? *Dissent*, 23 April. https://www.dissentmagazine.org/blog/whats-left-of-macondo (accessed 15 May 2017).

Benjamin, W. 1968. *Illuminations: Essays and reflections*. New York, Schocken.

Fanon, F. 2004. *The Wretched of the Earth*. Translated by Richard Philcox. New York, Grove Press.

García Márquez, G. 1970. *One Hundred Years of Solitude*. Translated by Gregory Rabassa. London, Penguin Books.

Gramsci, A. 1971. *Selections from the Prison Notebooks of Antonio Gramsci*. Translated by Quintin Hoare and Geoffrey Nowell Smith. New York, International Publishers.

Harvey, D. 2001. Globalization and the 'Spatial Fix', *Geographische Revue*, vol. 3, no. 2, pp. 23–30.

Hecht, S. B. and Cockburn, A. 1989. *The Fate of the Forest: Developers, destroyers and defenders of the Amazon*. London, Verso.

Hylton, F. 2007. Medellín's Makeover, *New Left Review*, vol. 44, pp. 71–89.

Hylton, F. 2006. *Evil Hour in Colombia*. London, Verso.

Lefebvre, H. 1991. *The Production of Space*. Oxford, Wiley-Blackwell.

Lutz, C. and Nonini, D. 1999. The economies of violence and the violence of economies, in H. L. Moore (ed.), *Anthropological Theory Today*. Cambridge, UK, Polity Press, pp. 73–113.

Marx, K. 1990. *Capital: A critique of political economy*, Vol. I. New York, Penguin Classics.

Mitchell, T. 1991. The limits of the state: beyond statist approaches and their critics, *American Political Science Review*, vol. 85, no. 1, pp. 77–96.

Parsons, J. J. 1967. *Antioquia's Corridor to the Sea: An historical geography of the settlement of Urabá*. Berkeley, University of California Press.

Pratt, M. L. 2008. *Imperial Eyes: Travel writing and transculturation*, 2nd edn. London, Routledge.

Redclift, M. R. 2006. *Frontiers: Histories of civil society and nature*. Cambridge, MA, MIT Press.

Reyes, A. 2009. *Guerreros y campesinos: El despojo de la tierra en Colombia*. Bogotá, Grupo Editorial Norma.

Rose, N. 1999. *Powers of Freedom: Reframing political thought*. Cambridge, UK: Cambridge University Press.

Steiner, C. 2000. *Imaginación y poder: El encuentro del interior con la costa en Urabá, 1900–1960*. Medellín, Editorial Universidad de Antioquia.

Tsing, A. L. 2005. *Friction: An ethnography of global connection*. Princeton, NJ, Princeton University Press.

USAID. 2009. *Más inversión para el desarrollo alternativo: Work Plan 2009–2010*. Bogotá, ARD Inc. for USAID Colombia.

Watts, M. 1992. Space for everything (A Commentary), *Cultural Anthropology*, vol. 7, no. 1, pp. 115–129.

Watts, M. 2004. Resource curse? Governmentality, oil and power in the Niger Delta, Nigeria, *Geopolitics*, vol. 9, no. 1, pp. 50–80.

Watts, M. 2006. The sinister political life of community: economies of violence and governable spaces in the Niger Delta, Nigeria, in G. W. Creed (ed.), *The Seductions of Community: Emancipations, Oppressions, Quandaries*. Santa Fe, NM: School of American Research Press, pp. 101–142.

Watts, M. 2012. A tale of two gulfs: life, death, and dispossession along two oil frontiers, *American Quarterly*, vol. 64, no. 3, pp. 437–467.

Watts, M. 2013. *Silent Violence: Food, famine, and peasantry in Northern Nigeria*. Athens, University of Georgia Press.

Williams, R. 1973. *The Country and the City*. Oxford UK, Oxford University Press.

12

Vibrancy of Refuse, Piety of Refusal

Infrastructures of Discard in Dakar[1]

Rosalind Fredericks

Garbage in Dakar is stinky, visceral matter. In addition to the ubiquitous plastic bags, household trash oozes with hard-to-manage organic matter like veggie peels, fish bones and animal guts. In the heat of the Senegalese sun, the remains of the day become a dangerous cauldron of stench, risk and stigma that is shouldered unevenly by residents across the city. And yet, in the right place and hands, garbage takes on a second life as sustenance and livelihood. What's more, in yet another context, household waste is wielded as a powerful weapon to call attention to the injustices of urban citizenship. Given its rich material-semiotic properties, it's no accident that trash has been a key feature of the distribution of precarity in neoliberal Dakar as well as the means by which urban residents have pioneered demands for a more just city. Over these last 25 years, garbage in the public space has manifested political battles at the heart of urban citizenship. Dakar's city streets have oscillated between remarkably tidy and dangerously insalubrious as the city's garbage infrastructure has become the symbol, object and stage for struggles over government, the value of labour and the dignity of the working poor.

Violence of Disposability

> The neoliberal tsunami broke with a dreadful ferocity on African cities … Reform – the privatization of public utilities creating massive corporate profits and a decline in service provision, the slashing

Other Geographies: The Influences Of Michael Watts, First Edition. Edited by Sharad Chari, Susanne Freidberg, Vinay Gidwani, Jesse Ribot and Wendy Wolford.
© 2017 John Wiley & Sons Ltd. Published 2017 by John Wiley & Sons Ltd.

of urban services, the immiseration of many sectors of the public
workforce, the collapse of manufactures and real wages, and often
the disappearance of the middle class – was remorselessly anti-urban
in its effects.

Watts 2006b, 6

The environments in which violence occurs must be construed
broadly to encompass a panoply of environmental uses and forms
of extraction, conservation, and rehabilitation ... Petroleum, soil
depletion, and tropical forest all possess quite different properties
and commodity characteristics – aside from their strategic or eco-
nomic value – that may, and often do, play a role in the dynamics
of violence or struggle.

Peluso and Watts 2001, 26

Violent neoliberal political economies congeal in Dakar's infrastructural
landscape. Radically uneven, sporadic and performative investments in
urban infrastructure have left parts of the city to rot, rust and slowly
crumble with the passage of time while others are spiffed up and glossed
over with elite, world-class urban aesthetics. Shrinking funding for urban
public services has unleashed intense volatility as different governing
bodies and politicians fight over diminishing budgets. For Dakar's garbage infrastructure, Senegal's neoliberal experiments have left much of the
collection equipment degraded, especially the system's backbone, the
garbage trucks, most of which are already imported used from Europe.
For ordinary *Dakarois*, this has meant increasingly challenging burdens
of managing waste in the home, already overwhelmingly shouldered by
household women. The poorest neighbourhoods pay the most for garbage
collection and become the most encumbered by garbage and the noxious
dangers of waste, pollution and disease. For the city's fifteen hundred or
so municipal trash workers, concerted efforts to flexibilize contracts and
scale-back benefits have meant increasingly precarious working conditions. They have found themselves at the centre of scathing political
battles to control this last bastion of the urban civil service. Garbage work
is messy business but garbage work without a contract or health care at
the mercy of relentless political manoeuvring can be downright perilous.

Over last 25 years there were over 12 major shake-ups of the institutions responsible for managing garbage, ranging from privatization of the
sector to full nationalization and a range of hybrid institutional forms
between the two (Fredericks 2013). A closer look at these reconfigurations
challenges broad-brush theories of neoliberal development. In contrast
with hegemonic representations of neoliberalism in African cities (Parnell
and Robinson 2012), reform in Dakar's garbage sector has triggered

widespread institutional tinkering, heightened state centralization and reinforced patronage modes of governing. Austerity has precipitated a mode of governing through disposability that brings people more firmly into the management of their own wastes. Governing through disposability involves governing through the distribution of precarity via the uneven provision of waste infrastructures and differentially disciplining people through waste in the production of so-called 'participatory citizenship'.

This latter point deserves unpacking. The imperative to shrink city budgets and scale back urban services in the era of austerity has turned on devolving urban infrastructure onto flexibilized labour. This has been accomplished through discourses of participation and the instrumentalization of community as a technology of development. A foundational moment of governing through disposability in the founding of the contemporary municipal garbage system in the social movement *Set/Setal*. Now Senegal's most famous youth social movement, *Set/Setal* ('Be Clean/ Make Clean' in Wolof) emerged in the wake of the contentious elections of 1988. In the context of political crisis and the fall-out from structural adjustment, *Set/Setal* youth aimed to cleanse the city in a moral sense and through literally removing wastes and beautifying public space (Diouf 1996; ENDA 1991). In the early 1990s, the Mayor of Dakar orchestrated the complete replacement of well-paid unionized garbage labourers for youth from the rowdy social movement through creating a city-wide participatory trash collection system. The new system served to flexibilize the labour force because the youth were contract labourers paid day-labour rates. In doing so, it allowed the Mayor to significantly slash the city budget and elide a labour dispute, all the while glorifying the new system in the name of participation.

In political terms, the tapping of *Set/Setal* was part of a new state strategy to battle for legitimacy while implementing urban public service reform. It functioned to quash what could have been the revolutionary potential of the youth activists and quell public dissent at structural adjustment and instead channel youth's energies towards the material act of cleaning the city. This patched the cracks in the state's hegemonic project and rejuvenated its image through making these young men and women the new face of the nation's orderly development.

The World Bank was a key partner in this transition. Besieged by criticism of the calamitous social consequences of loan conditionalities, the World Bank's new kinder, gentler 'revisionist' approach to austerity complemented reform with more attention to social safety nets (Mohan & Stokke 2000; Peck & Tickell 2002). The youth trash sector was soon managed by a new World Bank-funded public works agency which was rolled out in the 1990s with the goal of generating a significant number of low-skilled temporary jobs for unemployed youth and, in so doing, keeping the social peace (World Bank, 1997). In this new approach, the bonds formed during *Set/Setal* and youth's enthusiastic, entrepreneurial commitment to bettering their city became the 'stuff' to be instrumentalized by development (Watts 2006a). This was the beginning of the 'meteoric rise' of social capital theory as part of the broader 'discovery' of culture in development, and the networks of trust and social solidarity – the 'glue' binding communities – became *the* key building block of new ways to manage the city (Watts 2006a). The recourse to 'culture' and community as key infrastructural ingredients facilitated the scaling back of other technical and material components and, thus, the devolvement of waste infrastructure onto labour.

Institutional volatility, steadily increasing workloads, stagnating salaries and minimal investment in the sector through most of the last 25 years have wreaked havoc on the lives and bodies of those charged with doing the dirty work. The workers have borne the brunt of this labour-intensive infrastructure through the onerous physical demands of the work itself, associated diseases and the stigma of labouring in filth. As social juniors often unable to graduate into adulthood through founding their own homes, these youth have been subject to an array of social violences. Like many urban youth across the continent in the neoliberal era, they exist in a state of 'waithood' (Honwana 2012), trapped in a world with no foreseeable or calculable future (Buggenhagen 2001). Added to this is the stigma of working in a degraded sector shot through with all manner of negative associations.

The materiality of the labour process matters deeply for the bodies of the workers doing the dirty work. In addition to social violences, trash workers suffer a range of material insults associated with their specific daily labour conditions. Generally lacking in protective clothing and proper equipment, they stand nakedly exposed to an array of harms and shoulder disproportionate burdens of disease. Garbage is messy business and its collection is a delicate, expert labour. Degrading technology serves as an increasingly rickety material scaffolding and requires incessant expert labours of salvage *bricolage*. Like *bricoleurs* all over the continent, Dakar's trash workers devise ingenious strategies for transforming someone else's rubbish infrastructure into their own laboratories of innovation (see Mavhunga 2013). They are infrastructure hackers, manipulating the system's steel architecture through fastening their own bodies to the trucks' dysfunctional steel parts. The bodies and the machines conform to each other's labour as the workers bolster and manipulate the ailing trucks. The violence of this intimacy is starkly written in scars on the collectors' bodies.

Technologies of Community

Communities are not always warm and fuzzy.
Watts 2005b, 105

The new labor process therefore converts the domestic arena ... into a terrain of explicit contest and struggle.
Carney and Watts 1990, 201

Since *Set/Setal* in the late 1980s, a series of different low-tech, participatory formulas for devolving garbage infrastructure onto labour have resculpted the spaces, values and material burdens of labour for young men and women across the city. These new cultural economies of development are bluntly illuminated by a rash of NGO-run community-based waste management projects that sprang up in the city's periphery in the early 2000s. The idea behind these projects was to enhance garbage collection in poor, hard-to-reach neighbourhoods through off-grid, small-is-beautiful community-based infrastructure development.[2] The central pillar of the project in the Yoff district was replacing municipal garbage truck collection with horse-drawn carts and the volunteer labour of neighbourhood women who were charged with collecting their neighbours' garbage in the name of empowerment. A user fee for service effectively doubled households' expenditures on waste management.

Run by a well-respected Senegalese NGO, the projects were celebrated as a cutting-edge green solution to the challenges of urban development that reinvigorated a more 'appropriate', 'traditional' technology and social system of organization (ENDA 1999; Gaye and Diallo 1997). In practice, though supposedly non-governmental, they served the governmental ambitions of the local customary authority. Through employing neighbourhood women as 'municipal housekeepers'[3] and extending their social reproductive duties into the neighbourhood space, the male neighbourhood customary leadership was able to advance its autonomy from the local state. This was inscribed in a long historical conflict between municipal authorities and local community associations battling to control the destiny of Yoff in the face of urban change. In this way, the project reinforced local social-power hierarchies through further entrenching women's connection to waste. Just as in the municipal system, the insidious power of the community projects drew from the materiality of waste. Provisioned with only minimal equipment, if any, the women participants collected garbage with their bare hands. The arduous physicality of the collection process joined with the symbolic violence of being associated with waste: parading through their neighbourhoods wearing the smelly remains of their neighbours' waste threatened women's health and standing in their communities.

For these participating women and their neighbours, moreover, the result of this recourse to private, indirect government[4] built on and reconfigured household economies. The user fee interfaced with family power dynamics to disproportionately burden women with the costs associated with maintaining the household. In certain instances, women's responsibility for these new financial burdens and their concomitant reduced spending power precipitated intense conflicts between men and women heads of household. The women volunteers found themselves unwittingly caught in the middle of this tension. In spite of their lofty rhetoric, just as new, advanced technologies reconfigure social relations, these projects show that gaps in infrastructure and the devolution to so-called traditional technologies operate as means of control.

And yet, these neoliberal experiments with community-based waste management in the city's periphery were short-lived, owing to a confluence of quiet strategies of resistance and the very material territoriality of waste itself. The cumulative impact of widespread refusal by households to pay the user fee and the immense garbage pile that grew from the system's disarticulation from the grid ended up attracting unwanted attention from nuisance animals and higher authorities. When circling birds threatened air traffic near the airport, the Federal Aviation commission shut the project down and Yoff reverted back to regular municipal (truck) collection.

By the mid-2000s, a whole set of explicit rebellions around trash across the city would call into question the notion of flexibilizing municipal trash labour altogether.

Piety of Refusal

> The labor process is the point of incubation of historical and cultural fields of power in which human agency (organized through different social forms: firms, the state, households, and so on) is expressed. In turn these fields constitute institutional and discursive fields of struggle.
>
> Peluso and Watts 2001, 29

> Islam acts as both an alternative political and economic platform to the state and a critical oppositional discourse.
>
> Watts 1996, 284

Workers and residents in Dakar have exerted their rights to urban citizenship through tactics aimed at unsettling the 'proper' function and significance of trash infrastructures. From 2000 to 2009, the trash workers went from being disorganized, invisible and stigmatized to one of the most mobilized, well-known and respected labour unions, indeed social movements, in contemporary Senegal. Most importantly, starting in the mid-2000s, the trash workers have periodically disturbed the ordering processes of governing through disposability through a series of multi-day, havoc-wreaking, general trash strikes. During this time, ordinary *Dakarois* have joined in the chorus of rebellion through disorder by organizing trash strikes of their own in response to government inaction. These neighbourhood 'trash revolts', as they are commonly described, involved concerted dumping by whole neighbourhoods of their house-hold garbage into public streets, squares and government property. Strikes by workers and public dumping by residents deploy the power of dirt to creatively subvert ordering paradigms. Through disrupting the orderly flow of waste out of the city and intentionally externalizing pri-vate trash into public spaces, they creatively manifest disorder and reveal the 'public secret' of waste (see Hawkins 2003). In doing so, they contest governing prerogatives and all of the associated dimensions of stigma and abjection implied by living and working in filth.

Trash strikes are effective because they demonstrate the value of work-ers' labour as it is withdrawn but also because the material-semiotic resonance of trash as waste makes it a particularly powerful matter of rebellion. The public secret of waste and its associated risks relies on a multitude of everyday intersecting forms of vigilance to keep it in its proper place. Given the propensity for quick putrefaction, the proper function of the solid waste system requires unrelenting daily evacuation out of homes, into the waste grid and finally to the city's dump in the outskirts of the city to maintain urban order.[5] Years of tinkering has evolved the collection process towards a system premised upon intimate, daily intersections between women household garbage managers and municipal garbage collectors who share a commitment to ridding the city of its collective effluvium. The withdrawal of those labours and the accumulation of waste in public spaces frustrate the pursuit of moder-nity (Moore 2009). The filth of rotting garbage out of place offends the senses, renders the city impure and calls for the resolution of associated risks with urgency. In a culture where cleanliness of body and soul is of deep spiritual import, acts of dirtying or cleaning public space are pro-foundly meaningful.

Workers personally and publicly frame their labour as an act of Muslim piety. Cleanliness is godliness, they argue, so those who work to cleanse Dakar should be valued appropriately. In this way, they crea-tively wage a 'refusal to be refuse'[6] which turns the stigma of trashwork on its head. Through accompanying their strikes with a savvy public relations campaign emphasizing the value of cleaning in Islam, the trashworkers have redefined their profession, earned widespread public support and forced the state to react. As a mode of piety, cleaning labour becomes a form of religious agency through which workers develop their personal capacities to endure suffering and persevere in the face of difficult conditions. It can be seen as a mode of *bricolage* and maintenance of the self in a landscape of disrepair or pollution. Framed as a collective resource, it becomes a performative practice which evokes a shared moral compass. Appeals to Islamic morality cast a sharp ethical critique of the state's role in the flexibilization of labour and provide a new language through which to validate a vital infrastructure.

Garbage activists have emerged as one of the most important social movements in postcolonial Senegal and have played a key role in critiquing Senegal's neoliberal development trajectory. During this time, they have managed to significantly ameliorate public views of waste work, as evidenced by the trash revolts and widespread expressions of respect for and solidarity with the workers. The union's critiques have resulted in some very concrete improvements: after years of lobbying, the trash union's collective bargaining agreement was finally signed in 2014 conferring formal contracts, higher salaries and healthcare benefits to the sector's workers. Though there remains significant room for improvement in the labour conditions of the city's trash and other urban workers, with these gains, the garbage sector has pioneered the reversal of austerity management trends and heralded the possibility of a new era of urban governance in Dakar.

Infrastructure's Political Address

People are infrastructure, but what sort of structure is emerging from its citadel of girders and cables?

Watts 2005a, 183

Infrastructure is a powerful tool for understanding African cities, but the tumultuous history of waste politics in Dakar calls for a new toolbox (Appel, Anand and Gupta 2015) for thinking about infrastructure's address. Waste in Dakar urges for an understanding of infrastructure not as a simple, inert technical supporting structure, but as an assemblage of material, social and affective elements.[7] Infrastructures are ecologies[8] that assemble a range of spatialized relationships between political economic imperatives, technologies, natural processes, forms of sociality, social meanings and modes of ritual action. Considering these sociotechnical assemblages relationally allows us to probe the intersections between human and non-human agencies, the concrete burdens placed on labouring bodies and communities and the everyday meanings and practices through which infrastructures become political. Put simply, infrastructures help to ask: how do the building blocks of the city 'flicker, corrode, and lurch' (Fennell 2015) in specific conjunctures and with what concrete effects for urban lives?

The material being organized by infrastructures is an active agent in the political negotiations they engender. Like other managed objects and commodities – for instance, oil (Watts 2009), water (Anand 2011), sewage (McFarlane 2008), carbon (Whitington 2016), electricity (von Schnitzler 2013) and asbestos (Gregson, Watkins and Calestani 2010) – household trash has its own unique, context-specific materiality and spatiality. Many features of the infrastructural politics of trash in Senegal stem from the special political valence of waste as the opposite of value and its 'force' as vibrant matter (Bennett 2010; Braun and Whatmore 2010; Hawkins 2006).[9] Waste indexes processes of abjection in specific spatial and historical contexts and gathers force in its performativity and animation (Gidwani and Reddy 2011; Hawkins 2006). Whether animated through performances of government or staged revolt, the vibrancy of refuse in Dakar derives from the viscerality of its emplacement or displacement. The spatialities of discard and salvage *bricolage* build on and consolidate communicative channels and neighbourhood intimacy, and, in turn, enable the political possibilities that have emerged through striking and trash revolts. The force of these rebellions stems from the power of waste out of place (Douglas 1966).

Attention to waste in Dakar thus also foregrounds that urban infrastructures are composed of human as much as technical elements and that waste matters in its encounter with human bodies (Fredericks 2014). Even broader definitions of infrastructure in much of the recent critical literature often elide the way that new infrastructural assemblages are situated in human labour and the crucial intersections of human and non-human agencies. An interrogation of the ways that infrastructural ecologies assemble economies, matter, people and ideas allows for a more robust conception of *people as infrastructure* (Simone 2004) that responds to Watts' critique of what such a concept can provide for understanding and building more just African cities (Watts 2005a). An emphasis on labour shows how devolved, participatory waste infrastructures have come to be a central pillar of governing practices in Senegal and the material processes of abjection through which certain bodies become constituted as waste. The socio-spatial ordering of people and places is wrapped up with the symbolism of garbage and its material force as it intersects with human labour – or the emplacement of burdens of dirt and disease onto specific bodies and geographies. Waste infrastructures in Dakar thus illuminate the violences enacted within *bricolage* economies that predate their human elements but also highlight the important ways that infrastructures' people may upend these systems. Without rigorously attending to the *ways* that people are infrastructure, we risk ignoring the violences that may consolidate in the silences of infrastructure's concrete.

194   OTHER GEOGRAPHIES

Finally, this story draws attention to the intersections of materiality and social systems of meaning – or the generative capacity of non-human actants 'to move us and shape our collective attachments' (Braun and Whatmore 2010, xxiv, drawing on Connolly 2010). The material-semiotic qualities of waste are central not just to how these infrastructures operate, but to the structures of feeling to which they give rise. Waste's powers to disrupt and the salience of its opposite, cleanliness, as a symbol of faith and piety, are key features of the political valence of trash as vibrant matter in Dakar. The counterhegemonic force of trash rebellion in Dakar was forged out of the specific subjectivities conditioned by the corporeal practice of cleaning and manifested in the creative animation of the material itself in rebellion.

The piety of refusal raises important questions for the horizon of politics in urban Africa. Though there is a growing literature on ritual practices engaged by infrastructures and the spectral modes through which they are understood and valued, very little work has examined how this shapes concrete political action and citizenship claims. Much of the Africanist literature interprets spiritual understandings (particularly occult imaginings) of infrastructure as reactionary critiques of capitalism, globalization, neoliberalism and other elements of contemporary modernity.[10] But as Ishii points out, reducing spiritual understandings of infrastructure to modes of venting anxieties about modernity is inadequate for fully grappling with the rich 'ways in which [people] are entangled with, or encompassed by, nature and divinities' (2016, 4). Much more than simple moral panic at neoliberalism's violences, the piety of refusal represents a constructive striving to align moral and material economies in the wake of the failures of the secular nationalist project. It illuminates, moreover, the critical understandings that may be mobilized through religious identity or the 'dialectics of political and spiritual agency' (Diouf and Leichtman 2009, 12).[11] 'Modes of religiosity' forge new spaces of affiliation, movements, civic culture, and communities' (2009, 3–4) but also material and sociocultural infrastructures through which new moral geographies are crafted. Infrastructures require belonging; they are embedded in social relations and are erected upon moral architectures. Architectures of faith may be the 'citadel of girders and cables' through which more just political ecologies of infrastructures can be built.

Notes

1   The concerns taken up in this chapter are explored more fully in my book manuscript (Fredericks, forthcoming).
2   These are discussed more fully in Fredericks (2012, 2015).

3   Though with different governmental implications, this resonates with Miraftab's analysis of the manner in which neighbourhood women were employed as 'municipal housekeepers' in Cape Town (2004).
4   This can be seen as part of what Mbembe observed as a general trend in African development whereby private entities, including NGOs, increasingly occupy governmental space and control individual conduct (2001).
5   This of course is only the beginning of the second life of garbage. A village of people live at the city's dump, Mbeubeuss, carving lives (and value) out of the remains left by their better-off neighbours.
6   See Chari's similar notion of 'refusal to be detritus' (2013, 133).
7   There has been a recent explosion of literature arguing for broader conceptions of the social and political life of infrastructure. Some relevant overview pieces include (Appel et al. 2015; Graham and McFarlane 2015; Jensen and Morita 2016; Larkin 2013; McFarlane and Rutherford 2008; Mitchell 2014; O'Neill and Rodgers 2012; Simone 2015; Star 1999).
8   See Star (1999) and Murphy (2013).
9   In considering the political valence of vibrant matter, I am building on geographers' critiques of Bennett to move beyond an abstract, philosophical conception of politics (Braun et al. 2011; Castree 2011). This study is concerned with the imbalances of power between different actants and the concrete political work that non-human actants do in their intersection with human agencies.
10 For instance, see Masquelier (2002).
11 In contrast with the preoccupation within Senegalese religious studies of the role of the brotherhoods, rise of Islamism or the engagement of religious guides in electoral politics, Diouf and Leichtman attend to quotidian expressions of religiosity. See also Babou (2007) and Diouf (2013).

References

Anand, N. 2011. PRESSURE: the politechnics of water supply in Mumbai, *Cultural Anthropology*, vol. 26, no. 4, pp. 542–564.
Appel, H., Anand, N. and Gupta, A. 2015. Introduction: the infrastructure toolbox. from http://www.culanth.org/fieldsights/714-introduction-the-infrastructure-toolbox (accessed 15 May 2017).
Babou, C. A. 2007. Urbanizing Mystical Islam: Making Murid Space in the Cities of Senegal. *The International Journal of African Historical Studies*, vol. 40, *no.* 2, pp. 197–223.
Bennett, J. 2010. *Vibrant Matter: A political ecology of things*. Durham, NC, Duke University Press.
Braun, B., Anderson, B., Hinchliffe, S., Abrahamsson, C., Gregson, N. and Bennett, J. 2011. Book review forum: Vibrant Matter: A Political Ecology of Things, *Dialogues in Human Geography*, vol. 1, no. 3, pp. 390–406.
Braun, B. and Whatmore, S. J. (eds). 2010. *Political Matter: Technoscience, democracy, and public life*. Minneapolis: University of Minnesota Press.

Buggenhagen, B. A. 2001. Prophets and profits: gendered and generational visions of wealth and value in Senegalese Murid households, *Journal of Religion in Africa*, vol. XXXI, no. 4, pp. 373–401.

Carney, J. and Watts, M. 1990. Manufacturing dissent: work, gender and the politics of meaning in a peasant society, *Africa*, vol. 60, no. 2, pp. 205–241.

Castree, N. (2011). Jane Bennett, Vibrant Matter (Review), *Society and Space*, 14 September.

Chari, S. 2013. Detritus in Durban: polluted environs and the biopolitics of refusal, in L. A. Stoler (ed.), *Imperial Debris: On ruins and ruination*. Durham, NC, Duke University Press, pp.131–161.

Diouf, M. 1996. Urban youth and Senegalese politics: Dakar 1988–1994, *Public Culture*, vol. 8, pp. 225–249.

Diouf, M. (ed.). 2013. *Tolerance, Democracy, and Sufis in Senegal*. New York, Columbia University Press.

Diouf, M. and Leichtman, M. (eds.). 2009. *New Perspectives on Islam in Senegal*. New York, Palgrave Macmillan.

Douglas, M. 1966. *Purity and Danger: An analysis of the concepts of pollution and taboo*. London, Routledge.

ENDA. 1991. *Set Setal, des murs qui parlent: Nouvelle culture urbaine à Dakar*. Dakar, ENDA Tiers Monde.

ENDA. 1999. *Volet collecte des déchets et assainissement du quartier traditionnel de Yoff-Tonghor: Etude de faisabilité*. Dakar, Enda Tiers Monde R.U.P., République du Sénégal Commune d'Arrondissement de Yoff.

Fennell, C. 2015. Emplacement *Cultural Anthropology website, Theorizing the Contemporary*.

Fredericks, R. 2012. Valuing the dirty work: gendered trashwork in participatory Dakar, in C. Alexander and J. Reno (eds), *Economies of Recycling*. London, Zed Books, pp.119–142.

Fredericks, R. 2013. Disorderly Dakar: the politics of garbage in Senegal's capital city. *Journal of Modern African Studies*, vol. 51, no. 3, pp. 435–458.

Fredericks, R. 2014. Vital infrastructures of trash in Dakar, *Comparative Studies of South Asia, Africa, and the Middle East, Special Issue on Comparative Infrastructures*, vol. 34, no. 3.

Fredericks, R. 2015. Dirty work in the city: garbage and the crisis of social reproduction in Dakar, in K. Meehan and K. Strauss (eds.), *Precarious Worlds: Contested geographies of social reproduction*. Athens, University of Georgia Press, pp. 139–155.

Fredericks, R. forthcoming. *Garbage Citizenship: Vibrant infrastructures of discard in Dakar, Senegal*. Durham, NC, Duke University Press.

Gaye, M. and Diallo, F. 1997. Community participation in the management of the urban environment in Rufisque (Senegal), *Environment and Urbanization*, vol. 9, no. 1, pp. 9–29.

Gidwani, V. and Reddy, R. N. 2011. The afterlives of 'waste': notes from India for a minor history of capitalist surplus, *Antipode*, vol. 43, no. 5, pp. 1625–1658.

Graham, S. and McFarlane, C. (eds). 2015. *Infrastructural Lives: Urban infrastructure in context*. London, Routledge.

Gregson, N., Watkins, H. and Calestani, M. 2010. Inextinguishable fibres: demolition and the vital materialisms of asbestos, *Environment and Planning A*, vol. 42, no. 5, pp. 1065–1083.

Hawkins, G. 2003. Down the drain: shit and the politics of disturbance, in G. Hawkins and S. Muecke (eds.), *Culture and Waste: The creation and destruction of value*. Lanham, MD, Rowman and Littlefield Publishers, pp. 39–52.

Hawkins, G. 2006. *The Ethics of Waste*. Lanham, MD, Rowman and Littlefield.

Honwana, A. 2012. *The Time of Youth: Work, social change, and politics in Africa*. Sterling, VA, Kumarian Press.

Ishii, M. 2016. Caring for divine infrastructures: nature and spirits in a special economic zone in India, *Ethnos*, vol. 81, no. 2.

Jensen, C. B. and Morita, A. 2016. Infrastructures as ontological experiments (special issue) Ethnos, doi: 10.1080/00141844.2015.1107607

Larkin, B. 2013. The politics and poetics of infrastructure, *Annual Review of Anthropology*, vol. 42, pp. 327–343.

Masquelier, A. 2002. Road mythographies: space, mobility, and the historical imagination in postcolonial Niger, *American Ethnologist*, vol. 29, no. 4, pp. 829–856.

Mavhunga, C. C. 2013. What is Africa in technology? What is technology in Africa? (Keynote), MIT-Africa Interest Group, 1 October 2013. Cambridge, MA.

Mbembe, A. 2001. *On the Postcolony*. Berkeley, University of California Press.

McFarlane, C. 2008. Sanitation in Mumbai's informal settlements: state, 'slum', and infrastructure, *Environment and Planning A*, vol. 40, pp. 88–107.

McFarlane, C. and Rutherford, J. 2008. Political infrastructures (symposium), *International Journal of Urban and Regional Research*, vol. 32, no. 2.

Miraftab, F. 2004. Neoliberalism and casualization of public sector services: the case of waste collection services in Cape Town, South Africa, *International Journal of Urban and Regional Research*, vol. 28, no. 4, pp. 874–892.

Mitchell, T. 2014. Introduction, *Comparative Studies of South Asia, Africa, and the Middle East*, vol. 34, no. 3.

Mohan, G. and Stokke, K. 2000. Participatory development and empowerment: the dangers of localism, *Third World Quarterly*, vol. 20, pp. 247–268.

Moore, S. A. 2009. The excess of modernity: garbage politics in Oaxaca, Mexico, *The Professional Geographer*, vol. 61, no. 4, pp. 426–437.

Murphy, M. 2013. Chemical infrastructures of the St. Clair River, in S. Boudia and N. Jas (eds), *Toxicants, Health, and Regulation Since 1945*. London, Pickering and Chatto, pp. 103–115.

O'Neill, B. and Rodgers, D. 2012. Infrastructural violence (special issue), *Ethnography*, vol. 13, no. 4.

Parnell, S. and Robinson, J. 2012. (Re)Theorizing cities from the Global South: looking beyond neoliberalism, *Urban Geography*, vol. 33, no. 4, pp. 593–617.

Peck, J. and Tickell, A. 2002. Neoliberalizing space, *Antipode*, vol. 34, pp. 380–404.

Peluso, N. L. and Watts, M. (eds). 2001. *Violent Environments*. Ithaca, NY, Cornell University Press.

Simone, A. 2004. People as infrastructure: people as intersecting fragments in Johannesburg, *Public Culture*, vol. 16, no. 3, pp. 407–429.

Simone, A. 2015. Relational infrastructures in postcolonial urban worlds. In S. Graham and C. McFarlane (eds), *Infrastructural Lives: Urban Infrastructure in Context*. London, Routledge.

Star, S. 1999. The Ethnography of Infrastructure. *American Behavioral Scientist*, vol. 43, no. 3, pp. 377–391.

von Schnitzler, A. 2013. Traveling technologies: infrastructure, ethical regimes, and the materiality of politics in South Africa, *Cultural Anthropology*, vol. 28, no. 4, pp. 670–693.

Watts, M. 1996. Islamic modernities? Citizenship, civil society, and Islamism in a Nigerian city, *Public Culture*, vol. 8, pp. 251–289.

Watts, M. 2005a. Baudelaire over Berea, Simmel over Sandton? *Public Culture*, vol. 17, no. 1, pp. 181–192.

Watts, M. 2005b. The Sinister political life of community: economies of violence and governable spaces in the Niger Delta, Nigeria, in G. Creed (ed.), *The Romance of Community*. Santa Fe, NM, SAR Press, pp. 101–142.

Watts, M. 2006a. Culture, development, and global neo-liberalism, in S. A. Radcliffe (ed.), *Culture and Development in a Globalizing World: Geographies, actors, and paradigms*. New York, Routledge, pp. 30–57.

Watts, M. 2006b. Empire of oil: capitalist dispossession and the scramble for Africa, *Monthly Review*, vol. 58, no. 4, pp. 1–17.

Watts, M. 2009. The rule of oil: petro-politics and the anatomy of an insurgency, *Journal of African Development*, vol. 11, no. 2, pp. 27–56.

Whitington, J. 2016. Carbon as a metric of the human, *PoLAR: Political and Legal Anthropology Review*, vol. 39, no. 1, pp. 46–63.

World Bank. 1997. *Performance Audit Report (PAR) for the First and Second Public Works and Employment Projects (Agetip) (Republic of Senegal)* (No. 16516). Washington, DC, World Bank, Operations and Evaluations Department.

13

Reclamation, Displacement and Resiliency in Phnom Penh

Erin Collins

'The history of Phnom Penh is, in a sense, a history of water.'
Vann Molyvann (2003, 104)

'Man is the most important geographical reality of the delta.'
Pierre Gourou (1955, 109)

'Particular representations emerge from specific material spatial practices and from certain forms of domination and control of space, yet they can become material forces in their own right.'
Michael Watts (1992, 118)

There stands in central Phnom Penh, a 90-hectare expanse of tamped earth where the city's largest lake used to be. As with any landscape of erasure, the vast empty building plot belies a lived history of social re/production, struggle and dispossession. In 2006, Phnom Penh's greater Boeung Kak Lake area (see Figure 13.1) was passed over in the course of Cambodia's systematic land titling process. This omission in the official cadaster left more than four thousand households without title, many of whom had lived along the lake's shoreline since the late 1980s or early 1990s (Khemro and Payne 2004). These residents were categorized as squatters living on State Public Land.[1] Shortly thereafter, Cambodia's Council of Ministers approved the 99-year lease of the area to the private company of ruling party senator Lao Meng Khin, at the exceptional

Other Geographies: The Influences Of Michael Watts, First Edition. Edited by Sharad Chari, Susanne Freidberg, Vinay Gidwani, Jesse Ribot and Wendy Wolford.
© 2017 John Wiley & Sons Ltd. Published 2017 by John Wiley & Sons Ltd.

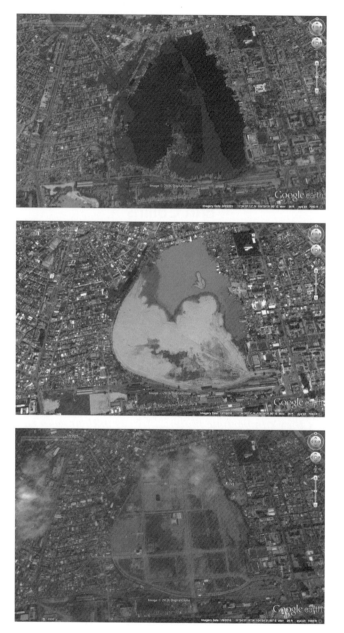

Figure 13.1   Google earth images of Boeung Kak Lake in 2003, 2010, 2016.

bargain of $.60 per square metre, at a time when such prime real estate was selling for between $700 and $1000 per square metre (Hughes 2008, 70). Following this non- competitive process, Lao Meng Khin's Shukaku Inc. entered into a joint venture with Erdos Hong Jun Investment Co.,

a Chinese firm from Inner Mongolia to develop the lake and surrounding area into a 'high-end residential, commercial and tourism
complex' (Yat and Shi 2014, 3). Shukaku Inc. began filling the lake with
sand in August 2008, completing the process in 2013.[2] As a 'lake of
natural origin' with 'public interest' Boeung Kak Lake was ostensibly
protected under Cambodia's 2001 Land Law (Grimsditch, Henderson,
Bugalski and Pred 2009). However, by filling the lake with sand Shukaku
Inc. transformed the lake into the far more flexible category of 'open land'.
Six months later, Prime Minister Hun Sen retroactively re-categorized the
area as State Private Land by sub-decree. In short, the material transformation of water into land was integral to both the political imagination
and the bureaucratic alchemy of the Cambodian state.

The scale of the Boeung Kak Lake project, as well as the brazenness of
its enclosure, has made land reclamation particularly prominent in current discussions of Phnom Penh's urban land politics. Yet, it is not at all
new. In this chapter I provide a historical tracing of this process, showing
how land reclamation has been a key means of rendering central urban
land alienable and its attendant populations displaceable from the colonial period to the present. I argue that land reclamation quite literally
contours urban precarity within the city – shoring up power and privilege while destabilizing the spatial claims of the urban poor. The chapter
proceeds as follows: I first root land reclamation in the colonial production of race and nature. I then trace it through biopolitical projects of
post-colonial and post-socialist urban remaking. I conclude with the
2012–2014 period of renewed intensive urban land conflicts, a moment
when land reclamation was increasingly articulated with and through
notions of resilience. Here I engage Watts' (2014) critique of 'resiliency
as a way of life', in which he offers a genealogy of resilience rooted in the
problem space of the Sahel desert.

Resilience, Watts writes, 'draws its strength from a historically distinctive sort of biopolitical apparatus … from powerful lines of … cata-
strophic thinking' (150). In many ways Phnom Penh is the urban
analogue of Watts' Sahel case. Phnom Penh is also a place where 'poverty
in the Global South meets up with global climate change' (155). Like the
Sahel, Phnom Penh was a key site of colonial experimentation (Wright
1991; Rabinow 1989). It is a place that looms large in the global imaginary of violence and deprivation. As the proxy theatre for the Vietnam
War, and the city that infamously fell to the Khmer Rouge on 17 April
1975, Phnom Penh is mimetically associated with war, trauma and genocide. More recently, a series of dramatic weather events lent Cambodia
the dubious honour of being named in 2013 the second most affected
country by climate change (Global Climate Risk Index 2015). In turn,
Phnom Penh was named as one of the Rockefeller Foundation's '100
Resilient Cities' in 2014. Ultimately, it is how Phnom Penh's geographic

features (low lying delta region, prone to flooding), social features (high poverty, high land insecurity, radical and rising inequality) and political features (seat of authoritarian state, endemic corruption) became articulated as a single, if shifting, milieu that interest me here. And second, how this history inflects contemporary struggles over land and power in the city.

In Phnom Penh, resilience is distinctly disjointed. This disjointedness is in part due to the ad hoc nature of governance in the time of climate change (Braun 2014, 51) to be sure. But it is also, I contend, a matter of imperial debris (Stoler 2008), exemplary of the 'differentially ordered material terrain in which past and present are unequally lived' (Chari 2013, 135). This chapter provides an account of how the afterlives of empire interfere as well as cohere within the politics of the present. In so doing I seek to contribute to recent scholarship that theorizes the complex and historically particular articulations of delta ecologies, urban infrastructure and forms of governance (CONS 2016; Ranganathan 2014, 2015; Morita and Jensen, forthcoming; Doyle 2012; Biggs 2012; Miller 2003; D'Souza 2006; Kinder 2015). In the course of this discussion, I work with the able tools of global political ecology and a geographical engagement with biopolitics, two bodies of thought fundamentally reconfigured by the intellectual force of Michael Watts (see Watts, 1983, 2008, 2012, 2014; Peet, Robbins and Watts 2010).

Land Reclamation and Racial Quarterization in Colonial Phnom Penh

Phnom Penh lies in a patchwork of small ponds, lakes and low-lying fields at the lower reaches of the Mekong River delta. The city clusters around the confluence of the Mekong, the Tonle Sap and the Bassac Rivers, making it optimally situated for trade and prone to flooding. According to traditional Southeast Asian conceptions of space and power, deltas are places where the land extended into the sea (Morita and Jensen, forthcoming, 3). Oriented to and by waterways, the metropoles of pre-colonial Southeast Asia were bound to each other across and via waterways, more than to their inland hinterlands. The predominance of oceanic modes of intercourse are partly credited for giving Southeast Asia its distinctive syncretic regional contours (Andaya 2006; Reid 2000; Tambiah 1977). These orientations were fundamentally reconfigured through colonial conquest. David Biggs recounts that what French colonial engineers disparaged as 'dead points' in the river – areas where the current stalled and silt accumulated – local people referred to as 'meeting points', (2012, xii) denoting two very different relationships to the delta ecology and its attendant socialites. French colonial officials applied themselves tirelessly to the spatial containment and technocratic management of fluctuations of water and sediment (Biggs 2012; Miller 2003) as did their counterparts in India (D'Souza 2006), and Thailand (Morita and Jensen, forthcoming). Civil engineering, it turned out, made for excellent social engineering.

In 1884 a visiting French journalist wrote of Phnom Penh's main road, 'Chinese, Malabars, Malay, Siamese, Annamites and Europeans live in the most abject promiscuity (Muller 2006, 61), highlighting the social fluidity that characterized the city in the early colonial period (Edwards 2007). Social and intimate forms of mixing were corrosive of the racial categories that secured colonial rule. Sexual intimacy was the source of considerable anxiety for the French colonial apparatus, manifest in a state obsession with mestissage children of mixed parentage (Stoler 1992). Meanwhile, the social intermixing of colonized subjects threatened the racialized division of labour that was essential to the colonial mode of production. The French placed Vietnamese in low-level bureaucratic positions throughout Indochina (Goscha 1995) and brought in Chinese labourers to work in rubber plantations (Edwards 2002).

Land reclamation was a key technology for the production of racialized difference in colonial Phnom Penh. In the 1890s the French dug a canal in a triangular pattern surrounding, and thus separating, the European quarter from the rest of the city. The earth dredged from the canal was used to infill swampy areas and to form the foundation of the streets of the French Quarter (Molyvann 2003, 47). Meanwhile the Khmer and Chinese quarters to the south of the city remained boggy, assembling the inter-referential elements of what one colonial functionary characterized as the '"Asiatic Milieu," watery-landscapes, fecundity, decay, populated by the "swamped race" of the Khmer' (Edwards 2007, 50). In Hebrard's Master Plan for Phnom Penh, realized in 1936,[3] the northern part of the city was demarcated as the French Quarter. Just South was the Chinese Quarter, and the Khmer population spilled out further south of the Chinese. The Chruy Chanva Peninsula, meanwhile, was slated for expansion of the French Quarter and included plans for new civic facilities. Though not explicitly demarcated on Hebrard's plan, the Vietnamese quarters were to the west of the Khmers' quarters and north of the Europeans, within the old Catholic village (Molyvann quoted in Evans, Falconer and Mills 2012, 14). By the late colonial period Phnom Penh's fluid topography had been sorted and secured within a matrix of racialized quarters (Wright 1991; Nam 2011).

By separating and shoring up the European Quarter from the broader inundated environs, colonial planners inscribed racial difference into the

infrastructural grid of the city. In so doing, they produced and then naturalized the urban ecology of the subaltern classes. And in so doing, racial
mixing was stemmed and racial hierarchy was spatialized. Of course, the
association of subalternity, inundated environs and precarity on the one
hand, and whiteness, dry land and security on the other, was – and is – a
matter of continuous construction. Imperial practices designed to shore
up human and ecological natures were from the beginning undermined
by the transitory materiality of the delta itself, engendering what Morita
and Jensen evocatively frame as 'two distinct and still open-ended
dynamics of their interlacing ontologies' (forthcoming, 4).

Land Reclamation as Nation Building Apparatus in the Post-Colonial City

Following independence in 1953, Prince Sihanouk briefly abdicated the
throne to preside over Cambodia's post-colonial nation building. A reso-
lute modernist as well as a prolific film maker, Sihanouk sought to refash-
ion Phnom Penh from a colonial city of bucolic charm to an engine and
showcase of Khmer modernity. Known as the Sankum Reastra Niyum
period (hereafter Sankum), this was a time of heightened nationalism and
intense creative production. The Sankum period was also a time of recombinant state racism. Ethnic Chinese and ethnic Vietnamese citizens were
banned from many jobs within the post-colonial state apparatus, even
while Sihanouk tapped these urban communities for monetary investment, political alliance and racial pageantry (Edwards 2002).

   The work of Cambodia's most famous architect and urbanist – Vann
Molyvann – is a lens onto the retooling of land reclamation as biopolitical apparatus in the post-colonial period. In 1957 Molyvann was
appointed the Chief Architect of the *Arrondissement des Batiments
Civils* and also served as the 'Architect/Advisor for the Municipality of
Phnom Penh' (Evans, Falconer and Mills 2012, 17). In this capacity,
Molyvann and his colleagues designed and built many of what remain
the city's most iconic buildings.[4] Trained in France in the modernist style
of Le Corbusier, Molyvann's vision incorporated elements of French
modern design, while also forging a distinctive Khmer aesthetic.
Molyvann and his colleagues made room for both the Khmer middle
class and water within the city. Molyvann declared 'water, natural light,
ventilation, and structural expression' to be 'the specific elements of a
Khmer ecological architecture'.[5]

   Molyvann's most extensive project was the building of the National
Stadium Complex in anticipation of the second GANEFO Games
(Games of the New Emerging Forces), along with an associated Athletes'

Village modelled on Corbusier's 'Unités d'Habitation' (Evans et al. 2012, 18). This project was of transnational as well as national renown. Three former collaborators of Le Corbusier, Vladimir Bodiansky, Guy Lemarchands and Gerald Hanning worked on the Athletes' Village project. Gerald Hanning was himself a central figure of the Athens Charter, a polemical document of urban planning that seized upon urban topography as object of social intervention.[6] The National Stadium was constructed on the site of a colonial era horse track, an existing Vietnamese settlement, and an 'insalubrious swamp'. The Athletes' Village, meanwhile, was sited along the Bassac Riverfront abutting the National Theatre. Reflecting on the site selection for the Athletes' village, Molyvann recalled:

> At the time, the area around Kab Ko [market] was an area of the city which often flooded. There were canals and a bridge which was called 'spean kon kat' (the mixed-blood child bridge). So the land on which we find the complex of buildings that today are known as 'buildings', as well as the land on which the National Theater stands, was all established by using silt dredged from the river to raise the whole area above the flood level. (Vann Molyvann 2001, 8)

Following development of the site, municipal officials resettled the former Vietnamese settlement on the outskirts of the city.[7] The Athletes' Village apartments, meanwhile, provided much needed affordable housing for Cambodian civil servants after the GANEFO Games.

Here then, we have the pillars of national identity – sport, culture and civil service – built on top of land 'reclaimed' from a racialized subaltern population, who were then displaced to the inundated environs of the urban periphery. While the intimate marker of 'bridge of the mixed-blood child' condenses several elements of the inundated milieu of the colonial era – the watery environs of the site, the taboo of miscegenation and racialized dispossession – land reclamation did not merely reproduce colonial categories in the post-colonial city. Rather, in this period land reclamation was honed into a more precise tool for the excision of specific subaltern populations. In place of the rigid spatial demarcation of colonial quarterization, land reclamation now corresponded to a more 'natural', 'vernacular' and indeed 'resilient' mode of urban planning, in which water was incorporated and managed within the city rather than held at bay.

Sihanouk's nation-building programme was cut short by a coup d'état in 1970. What followed was a tragic decade of escalating US bombing of the Cambodian countryside and peri-urban fringe and then the fall of Phnom Penh to Khmer Rouge forces in 1975. In the course of the short but horrendous Khmer Rouge era (1975–1979) nearly two million Cambodians (between a fifth and a quarter of the total population) were allowed to die or made to die (Kiernan 2014; Chandler 1999; Um 2015; Tyner 2008). The Khmer Rouge pursued many large-scale earthwork projects, though as they were predominantly sited in rural areas, they are outside of the scope of this discussion.

In January of 1979, a coalition of Cambodian resistance forces and Vietnamese troops liberated Cambodia from the Khmer Rouge regime, inaugurating the post-Khmer Rouge, socialist People's Republic of Kampuchea (PRK) decade of socialist reconstruction. The majority of those who settled in Phnom Penh in the 1980s were new to the city, and at first occupied empty homes according to a socialist urban planning regime oriented around labour and security (Collins, forthcoming). However, by the mid-1980s the city was full, and latecomers predominantly settled on open land, such as along railroad tracks and canals. In the 1980s ethnic Vietnamese and ethnic Cham fisher people in particular settled upon seasonally inundated lands along the riverfront. During the dry season, they lived in the city, departing on boats to fish along the Tonle Sap when the monsoon rains began. While the housed population in 1991 was estimated at 700 000, there were an estimated additional 100 000 and 200 000 such seasonal occupants of inundated land (Yap, Standley and Ottolenghi 1992, 21).

Accumulation by Land Reclamation in the Post-Socialist City

Cambodia's transitional period of the late 1980s and early 1990s entailed the development of a speculative land market and concomitant processes of dispossession. In April of 1989, the communist, PRK state changed their name to the State of Cambodia (SOC) and instituted a set of liberal reforms. Chief among these was the reintroduction of limited private property rights in urban areas and rights of possession in rural areas. Between April of 1989 and February of 1992, the SOC distributed 4.9 million receipts for private property claims made to land throughout the country.[8] Yet, title was not evenly bestowed. Land reform produced a new urban binary of landowners and squatters and yielded widespread dispossession throughout the city (Collins 2016a). The difference between dry land and seasonally inundated land was in many cases determinate of whether a parcel was allocated as private property or was retained as state land (Collins, 2016b). In this context land reclamation was integral to the making of post-socialist urban subjectivities of 'squatters' and 'owners'.

In this period, the designation of a property as state land did not necessarily preserve it as public domain. Rather, state land was the condition of emergence for high-end private developments of the politically well connected. The case file from a land dispute in Russey Keo District illustrates how this worked on the ground:

> Land dispute between a collective group of people and their Chief of Village who allegedly sold communal land and kept the money personally. The land in question contains a pond, the water of which is used by the people of the village communally, which is the properties' sole value to the people. To quiet the uproar of the 'sale' of this land, the pond is being surreptitiously filled in so that there is nothing left to argue about. The villages think this may be finished within two days. Can anything be done to enjoin the filling in of this pond pending the tribunal's decision on the merits of this case?[9]

Here the physical process of sand dredging wore away villagers' claims through the change in the material form of the property itself. By 'reclaiming' the land the chief, in fact, claimed the lake. He erased the claims of those already living around the pond, rendering it 'open land' that could then be carved into plots and sold. Working in and through each other, processes of material transformation and bureaucratic alchemy congealed Phnom Penh's radically uneven post-socialist urban political topography.

Concurrent with the sharp rise in urban land values, state records show an intensification of land reclamation and sand dredging in the city. A report from March 1991 describes an ongoing project of dredging sand from the Bassac River for use as infill of three urban developments.[10] Meanwhile, the People's Committee of Phnom Penh recommended the dredging of sand in the Tonle Bassac River to infill new development zones in Russei Keo and Toul Baikong. Two months later, the Cabinet Council of Ministers authorized the pumping of sand to enlarge the city.[11]

Capitalist accumulation by land reclamation worked through channels carved by colonial and nationalist urban planning regimes. Yet, it took on new dynamics in the transitional period. It is in this period that resilience as dynamic systemic adaptation – the city's capacity to absorb and disperse seasonal floods across the urban body – ran aground on resilience as mode of self-fashioning. By filling in boggy urban environs, individual opportunists and the municipality more generally, transformed water into capital. 'At once a means of life and a mode of capture' (Braun 2014, 55), land reclamation was a technology well suited to 'the brave new world of turbulent capitalism and the global neoliberal order' (Watts 2014, 169).

Resiliency in the Time of Climate Change

Flying under the masthead of 'The Future of Resilience',[12] the 'Next City' blog locates Phnom Penh's 2010s-era endemic flooding at the confluence of corrupt building practices epitomized in Boeng Kak Lake, and a 'culture of throwing garbage into the drains' (Otis 2013)[13] This formulation conjures the caricatured villains of the 'Third World': corruption and a culture of filth. It is a time-honoured pairing that bears a distinctive colonial accent. And it is a representation that underscores Watts' claim that

> [r]esilience, as it has emerged as a set of practices deployed by state and civil society groups, forms the basis for addressing the uncertainties and instabilities not simply of nature, but of contemporary capitalism and the national security state, and it does so by endorsing a distinctive form of biopolitics and technologies of the self. (2014, 149–150)

In 2008 sand dredging for development in the central city reputedly caused an excess of seventy thousand cubic metres of land to slide into the Mekong in Phnom Penh's poor, outer district of Mean Chey (Channyda, 2008). That year, Mean Chey experienced the worst floods on record, and the area has fared nearly as badly in subsequent years (Becker and Prak 2009). In 2011, a thousand homes surrounding the former Boeung Kak Lake suffered such extensive damage that they had to be abandoned (Schneider 2011, 9). At the same time, city planners and foreign technical advisors see land reclamation, sand dredging and drainage infrastructure as essential technologies for the management of flood waters within the city – rendered all the more critical in the face of climate changed and development induced precarity.[14] A suite of NGOs currently operating in Phnom Penh's urban poor sector support informal communities to self-upgrade their drainage infrastructure as a means of leveraging urban improvements for municipal recognition and land rights. In other words, land reclamation is touted as both the *cause* of and the *solution* to flooding, urban precarity and ecological destruction in contemporary Phnom Penh.

A brief vignette from my fieldwork is suggestive of how this contradiction plays out within urban land politics. A number of informal communities in which I conducted interviews in 2012 and 2013 were involved in extensive self-financed drainage projects of their untitled settlements. A resident of one such settlement explained: 'We are pulling up the dirt streets and putting in drainage pipes, so that the area will not flood so badly. Through our actions we show the municipality that we are not anarchists, we care about the beauty of the city.'[15] Here then, a shining example of, a spontaneous market order ... built from and out of self-making and self-regulation by individuals and communities through calculation and commodification shaped by their own peculiar exposure to the necessary and unavoidable contingencies of life. (Watts 2014, 153)

Yet, this particular case also illustrates the ways in which resiliency as 'calculative metric' honed to the individual neoliberal subject interferes with resiliency as systemic and adaptive response to chronic and acute flooding. An NGO worker who worked closely with this community related that residents had resisted pooling money and resources to do the drainage work 'properly'. They had, he said, insisted on each ripping up their 'own stretch of street, laying their own drainage pipes', which he argued resulted in a more expensive and less effective drainage system composed of 'differently sized, poorly graded pipe'. Here resiliency as 'synoptic field theory' is disrupted by resiliency as mode of individual self-fashioning. Phnom Penh's urban poor have a long historical memory of land reclamation as a technology productive of uneven exposure to displacement in the always already unstable delta ecology. They are acutely aware that the flooding of their urban environs renders them particularly vulnerable to displacement, and go to great lengths to be seen as individual property owners rather than as massified subaltern subjects.

Contemporary projects of land reclamation that work with the fluctuations and fluidity of Phnom Penh's delta ecology are 'grafted onto', as Watts puts it, the critical (in this case vernacular, post-colonial and post-socialist) responses to colonial modes of governance (2014, 164). Yet, the particular contouring of the subaltern population through the apparatus of land reclamation – present in recombinant form through Phnom Penh's multiple, sequential urban planning regimes – destabilizes the very 'resilient' programmes and projects that work through it. Resilience as a way of life in contemporary Phnom Penh is exemplary of how 'particular representations emerge from specific material spatial practices and from certain forms of domination and control of space, yet they can become material forces in their own right' (Watts 1992, 118). Or, as put by Vann Molyvann, it is a story about how 'the history of Phnom Penh is, in a sense, the history of water' (2003, 104).

Notes

1   http://www.inclusivedevelopment.net/bkl/ (accessed 15 February 2014).
2   90% of the original surface area of the lake is now land.
3   See Jennings (2009) for a fascinating comparative discussion of Hebrard's 1920s master plans for Dalat, Hanoi and Phnom Penh.

4 Including the Independence Monument, the 100 houses project for civil servants, the Bassac Athletes' village and National Stadium Complex, the National Theatre and the Teacher's Training College.

5 The Agency Interview: The Litany of Power, the Legacy of Modernism. Excerpts from interviews conducted by Bill Greaves in Phnom Penh in 2010 printed in K. Evans, I. Falconer and I. Mills, 2012, Yale School of Architecture, and given to the author by Dr Vann Molyvann at the time of the interview. See the Vann Molyvann Project for a compendium of this important work of documentation: http://www.vannmolyvannproject.org/ (accessed 15 May 2017).

6 A key passage from the Athens Charter reads: 'The geographic and topographic situation is of prime importance, and includes natural elements, land and water, flora, soil, climate, etc ... Unsanitary slums should be demolished and replaced by open space. This would ameliorate the surrounding areas ... the demolition of slums surrounding historic monuments provides an opportunity to create new open spaces.' The Athens Charter (1933) was written at the IVth International Congress for Modern Architecture: 'the functional city' focused on urbanism and the importance of planning.

7 In this case, and in contrast to contemporary programs of dispossession, the municipality provided those displaced from the central city with plots of land in Mean Chey, in an area 'fully equipped with roads, water supply and drainage' (Molyvann 2001, 8).

8 Document Ref: AC/ER/DJB-16.92. UNTAC Dispute Files, National Archive of Cambodia, Phnom Penh.

9 File 3.057, UNTAC Dispute File, National Archive of Cambodia, Phnom Penh.

10 Boeung Snout lake, Phnom Penh Thmai and the Chrui Chanva Peninsula.

11 Document no. 75 SSR, dated 24 May 1991.

12 https://nextcity.org/daily/entry/bringing-japanese-engineering-to-phnom-penhs-beleaguered-drainage-system (accessed 15 May 2017).

13 Here Otis quotes Uchida Togo who oversees JICA's drainage and flood protection work in Cambodia.

14 See Project on Capacity Building for Urban Water Supply System in Cambodia (Phase 3)/2012.11-2017.11 [Phnom Penh, Pursat, Battambang, Siem Reap, Kompong Tom, Kompong Cham, Svay Rieng, Kampot and Sihanoukville]. http://www.jica.go.jp/cambodia/english/activities/index.html (accessed 15 August 2016).

15 Author's field notes, Phnom Penh, August 2013.

References

The Agency Interview: The Litany of Power, the Legacy of Modernism. 2012. Excerpts from interviews conducted by Bill Greaves in Phnom Penh in 2010 printed in K. Evans, I. Falconer and I. Mills. Yale School of Architecture.

Andaya, B. W. 2006. The Flaming Womb: Repositioning women in early modern Southeast Asia. Honolulu, University of Hawaii Press.

Becker, A. and Prak, C. T. 2009. Filling lake could cause flooding study finds, in *The Cambodia Daily*, 12 March.

Biggs, D. 2012. *Quagmire: Nation-Building and Nature in the Mekong Delta*. Seattle, WA, University of Washington Press.

Braun, B. P. 2014. A new urban dispositif? Governing life in an age of climate change, *Environment and Planning D: Society and Space*, vol. 32, no. 1, pp. 49–64.

Chandler, D. P. 1999. *Brother Number One: A political biography of Pol Pot*. New York, Westview Press.

Chari, S. 2013. Detritus in Durban: polluted environs and the biopolitics of refusal, in A. L. Stoler (ed.), *Imperial Debris: On Ruins and Ruination*. Durham, NC, Duke University Press, pp. 131–161.

Channyda, C. 2008. Water resources ministry lifts ban on sand dredging in river, in *The Cambodia Daily*, 14 January.

Collins, E. 2016a. Postsocialist informality: the making of owners, squatters and state rule in Phnom Penh, Cambodia (1989–1993), *Environment and Planning A*. vol. 48, no. 12, pp. 2367–2382.

Collins, E. 2016b. Repatriation, refoulement, repair. *Development and Change*, vol. 47, no. 6, pp. 1229–1246.

Collins, E. forthcoming. *Rupture as Chronopolitics in Phnom Penh*.

Cons, J. 2016. *Sensitive Space: Fragmented territory at the India-Bangladesh border*. Seattle, WA, University of Washington Press, 2016.

Doyle, S. E. 2012. City of water: architecture, urbanism and the floods of Phnom Penh. *Nakhara: Journal of Environmental Design and Planning*, vol. 8, pp. 135–154.

D'Souza, R. 2006. *Drowned and Dammed: Colonial Capitalism and Flood Control in Eastern India*. New Delhi, India, Oxford University Press.

Edwards, P. 2007. *Cambodge: The cultivation of a nation, 1860–1945*. Honolulu, University of Hawaii Press.

Edwards, P. 2002. Time travels: locating xinyimin in Sino-Cambodian histories, in P. Nyírí and I. Saveliev (eds), *Globalizing Chinese Migration: Trends in Europe and Asia*. Burlington, UK, Ashgate Press, pp. 254–290.

Evans, K., Falconer, I. and Mills, I. 2012. *Agency*, Yale School of Architecture. Cambridge, MA, MIT Press.

Foucault, M. 1980. The confession of the flesh, in C. Gordon (ed.), *Power/Knowledge: Selected Interviews and Other Writings*. New York, Pantheon Books, pp. 194–228.

Gourou, P. 1955. *The Peasants of the Tonkin Delta: A study of human geography* (Vol. 1). New Haven, CT, Human Relations Area Files.

Goscha, C. E. 1995. *Vietnam Or Indochina? Contesting Concepts of Space in Vietnamese Nationalism, 1887–1954* (Vol. 28). Copenhagen, Denmark, NIAS Books.

Grimsditch, M., Henderson, N., Bugalski, N. and Pred, D. 2009. *Untitled: Tenure insecurity and inequality in the Cambodian land sector*. Phnom Penh, Cambodia, Babsea, Cohre, JRS.

Hughes, C. 2008. Cambodia in 2007: development and dispossession, *Asian Survey*, vol. 48, no. 1, pp. 69–74.

Jennings, E. T. 2009. Đà Lạt, Capital of Indochina: remolding frameworks and spaces in the Late Colonial Era, *Journal of Vietnamese Studies*, vol. 4, no. 2. pp. 1–33.

Khemro, B. H. S. and Payne, G. 2004. Improving tenure security for the urban poor in Phnom Penh, Cambodia: an analytical case study, *Habitat International*, vol. 28, no. 2, pp. 181–201.

Kiernan, B. 2014. *The Pol Pot Regime: Race, power, and genocide in Cambodia under the Khmer Rouge, 1975–79*. New Haven, CT, Yale University Press.

Kinder, K. 2015. *The Politics of Urban Water: Changing waterscapes in Amsterdam*. Athens, University of Georgia Press.

Miller, F. 2003. *Society-Water Relations in the Mekong Delta: A political ecology of risk*. University of Sydney, Australia, Division of Geography, School of Geosciences.

Molyvann, V. 2003. *Modern Khmer Cities*. Phnom Penh, Reyum Publishing House.

Molyvann, V. 2001. *A Conversation with Vann Molyvann*. Phnom Penh, Reyum Publishing House.

Morita, A. and Bruun Jensen, C. forthcoming. Delta ontologies: infrastructural transformations in Southeast Asia. *Social Analysis*.

Muller, G. 2006. *Colonial Cambodia's 'Bad Frenchmen': The rise of French rule and the life of Thomas Caraman, 1840–87*. London, Routledge.

Nam, S. 2011. Phnom Penh: from the politics of ruin to the possibilities of return, *Traditional Dwellings and Settlements Review*, vol. 22, pp. 55–68.

Otis, D. 2013. *Can Japanese engineering fix Phnom Penh's creaky drainage infrastructure?* Next City blog post: https://nextcity.org/daily/entry/bringing-japanese-engineering-to-phnom-penhs-beleaguered-drainage-system on The Future of Resilience. 3 November 2013 (accessed 15 August 2016).

Peet, R., Robbins, P. and Watts, M. (eds). 2011. *Global Political Ecology*. London, Routledge.

Rabinow, P. 1989. *French Modern: Norms and forms of the social environment*. Cambridge, MA, MIT Press.

Ranganathan, M. 2014. Paying for pipes, claiming citizenship: political agency and water reforms at the urban periphery, *International Journal of Urban and Regional Research*, vol. 38, no. 2, pp. 590–608.

Ranganathan, M. 2015. Storm drains as assemblages: the political ecology of flood risk in post-colonial Bangalore, *Antipode*, vol. 47, no. 5, pp. 1300–1320.

Reid, A. 2000. *Charting the Shape of Early Modern Southeast Asia*. Chiang Mai, Thailand, Silkworm Books.

Schneider, H. 2011. The conflict for Boeng Kak lake in Phnom Penh, Cambodia. *Pacific News*, Vol. 36 (July/August), pp. 4, 10.

Stoler, A. 1992. Sexual affronts and racial frontiers: European identities and the cultural politics of exclusions in colonial Southeast Asia, *Comparative Studies in Society and History*, vol. 34, no. 3, pp. 514–551.

Stoler, A. 2008. Imperial debris: reflections on ruins and ruination, *Cultural Anthropology*, vol. 23, no. 2, pp. 191–219.

Tyner, J. A. 2008. *The Killing of Cambodia: Geography, genocide and the unmaking of space*. Aldershot, UK, Ashgate Publishing.

Tambiah, S. J. 1977. The galactic polity: the structure of traditional kingdoms in Southeast Asia, *Annals of the New York Academy of Sciences*, vol. 293, no. 1, pp. 69–97.

Um, K. 2015. *From the Land of Shadows: War, Revolution, and the Making of the Cambodian Diaspora*. New York and London, NYU Press. http://www. refworld.org/docid/5051ac562.html (accessed 12 March 2016).

Watts, M. 1983. On the poverty of theory: natural hazards research in context, in K. Hewitt (ed.), *Interpretation of Calamity: From the Viewpoint of Human Ecology*. Boston, MA, Allen & Unwin, pp. 231–262.

Watts, M. 1992. Space for everything (a commentary), *Cultural Anthropology*, vol. 7, no. 1, pp. 115–129.

Watts, M. 2013. *Silent Violence: Food, famine, and peasantry in northern Nigeria* (Vol. 15). Athens, University of Georgia Press.

Watts, M. 2014. Resilience as a way of life: biopolitical security, catastrophism, and the food–climate change question, in Nancy N. Chen and Lesley A. Sharp (eds), *Bioinsecurity and Vulnerability*, Santa Fe, NM, SAR Press, pp. 145–172.

Watts, M. and Kashi, E. 2012. *Curse of the Black Gold; 50 Years in the Niger Delta*. Brooklyn, NY, powerHouse Books.

Wright, G. 1991. *The Politics of Design in French Colonial Urbanism*. Chicago, University of Chicago Press.

Yap K. S., Standley, T. and Ottolenghi, R. (1992) Report of the UNCHS (HABITAT) Needs Assessment Mission in the Urban Sector in the Light of the Imminent Influx of Returnees in Cambodia, Unpublished report (25 April–24 May 1992).

Yat, Y. and Shi, Y. 2014. The failure of implementation of land legislations is a root cause of land disputes in Cambodia: case studies of Beoung Kak Lake and Kampong Speu Sugar Company, *International Journal of Interdisciplinary and Multidisciplinary Studies*, vol. 2, no. 1, pp. 1–8.

Index

Other Geographies: The Influences Of Michael Watts, First Edition. Edited by Sharad Chari,
Susanne Freidberg, Vinay Gidwani, Jesse Ribot and Wendy Wolford.
© 2017 John Wiley & Sons Ltd. Published 2017 by John Wiley & Sons Ltd.